THE
BIG STALL

HOW BIG OIL AND THINK TANKS ARE BLOCKING ACTION ON CLIMATE CHANGE IN CANADA

DONALD GUTSTEIN

JAMES LORIMER & COMPANY LTD., PUBLISHERS
TORONTO

James Lorimer & Company Ltd., Publishers acknowledges funding support from the Ontario Arts Council (OAC), an agency of the Government of Ontario. We acknowledge the support of the Canada Council for the Arts, which last year invested $153 million to bring the arts to Canadians throughout the country. This project has been made possible in part by the Government of Canada and with the support of the Ontario Media Development Corporation.

Cover design: Tyler Cleroux

Library and Archives Canada Cataloguing in Publication

Gutstein, Donald, 1938-, author
       The big stall : how big oil and think tanks are blocking
action on climate change in Canada / Donald Gutstein.

Includes bibliographical references and index.
Issued in print and electronic formats.
ISBN 978-1-4594-1347-4 (softcover).--ISBN 978-1-4594-1348-1 (EPUB)

       1. Petroleum industry and trade--Environmental aspects--
Canada. 2. Petroleum industry and trade--Political aspects--Canada.
3. Petroleum industry and trade--Economic aspects--Canada. 4. Fossil
fuels--Environmental aspects--Canada. 5. Fossil fuels--Political
aspects--Canada. 6. Fossil fuels--Economic aspects--Canada. 7. Global
warming--Canada. 8. Climatic changes--Canada. 9. Environmentalism--
Canada. 10. Energy industries--Canada--Forecasting. I. Title.

HD9574.C22G88 2018   338.2'7280971   C2018-904026-2
                                                    C2018-904027-0

James Lorimer & Company Ltd., Publishers
117 Peter Street, Suite 304
Toronto, ON, Canada
M5V 0M3
www.lorimer.ca

Printed and bound in Canada.

*[W]e must investigate any presentation of the truth
that claims to be authoritative.*
— Dzogchen Ponlop Rinpoche, *Rebel Buddha*

CONTENTS

ABBREVIATIONS

BCNI = Business Council on National Issues
BCSD = Business Council for Sustainable Development
CAPP = Canadian Association of Petroleum Producers
COP = Conference of Parties
EDF = Environmental Defense Fund
EPA = Environmental Protection Agency
EPIC = Energy Policy Institute of Canada
IETA = International Emissions Trading Association
IPCC = Intergovernmental Panel on Climate Change
NEP = National Energy Program
NGO = Non-governmental organization
OECD = Organization for Economic Cooperation and Development
OPEC = Organization of the Petroleum Exporting Countries
UNFCCC = United Nations Framework Convention on Climate Change
WBCSD = World Business Council for Sustainable Development

INTRODUCTION

For most of the world, the oil crisis of the 1970s and the signing of the Paris Climate Agreement in 2015 have little in common beyond the fact they both were all about humanity's seemingly bottomless appetite for burning fossil fuels.

But Canada has an additional commonality: A member of the Trudeau family was leading the country during each of these events.

Prime Minister Pierre Trudeau took dramatic action, creating a national energy company and exerting aggressive public oversight of the industry, in the process enraging the big oil companies and their allies in Edmonton and Washington, D.C.

His son, Prime Minister Justin Trudeau, put forward modest measures Big Oil itself had been advocating for a decade, receiving industry's plaudits.

One Trudeau tried to counter Big Oil's dominance; the other did Big Oil's bidding. Pierre Trudeau's message was this: Canadian oil policy must be for the benefit of Canadians. Justin Trudeau's message was this: Climate change isn't a crisis but a market opportunity. We can deal with it by putting a price on carbon and by investing in clean growth.

How did this happen? How did we go from giving the oil industry orders to having the oil industry dictate climate policy?

The answer could be simply that father and son had very different personalities and values. Or it could be that the oil crisis, with its immediate impacts on consumers, workers and investors, required more direct intervention than global warming, a problem that could be pushed off to the future.

But more was at play. When Trudeau the elder created Petro-Canada and introduced the National Energy Program, Keynesianism still reigned supreme. Government intervention in the economy was legitimate. By the time of Trudeau the younger, neoliberalism had transformed economic and political thinking, decreeing that only the market can make decisions.

Neoliberalism reduces the role of government to creating and enforcing markets, and propping them up when they fail, as in the 2008 financial meltdown. Otherwise, just get out of the way. From a fringe movement in the 1950s, neoliberalism rose to become our prevailing wisdom in the 1980s. It infiltrated the media, the economics profession and the political elite. Neoliberals understood that politicians may govern us, but ideas govern politicians. So whoever controls the ideas about how to address global warming will prevail. As Margaret Thatcher put it at the dawn of the neoliberal age, "there is no alternative"; forty years later, there still isn't.

Neoliberalism didn't just happen. It was spread through a network of seemingly independent think tanks that package neoliberal doctrine for delivery to media, politicians, policy-makers and the public through studies, reports, op-eds, websites and social media.

The effect of think tanks on public discourse has been a theme in my previous books, most recently *Harperism*, published in 2014. But it is in the current setting that think tanks have shown their corrosive power — how, in concert with a pliable media, they reduce the public interest to the interests of the 1 per cent.

Manufacturing doubt about the relationship between carbon dioxide emissions and climate change is one of neoliberalism's greatest triumphs. As of 2018, a sizeable minority of Canadians — particularly those in Western Canada and the Conservative Party — still doubted that global warming was real, which threatened any kind of government action.

This dumbing-down was achieved in several concerted campaigns of evasion.

At first, with the support of the think tank network, Big Oil denied climate change. Then, in a second delaying tactic, it switched to advocating voluntary actions by individual companies and carbon markets. Finally, in 2007, it decided on the way forward. Climate change was real, the industry acknowledged. But instead of being a threat to the bottom line it presented a golden opportunity to develop new markets. These were to be encouraged by establishing modest carbon dioxide emission reduction targets and modest carbon pricing, and promoting growth through government subsidies for clean technologies in a plan that included the approval of new pipelines.

As in each of the previous evasions, the industry had staved off a day of reckoning that would have reduced Canada's carbon footprint.

During 2018, Justin Trudeau faced mounting obstacles to the execution of the plan. It was called a "grand bargain" by industry: we'll agree to carbon pricing if you allow pipeline expansion. Trudeau was having difficulty delivering on either side of the equation. Some provinces dragged their heels on any kind of pricing scheme — with the election of Doug Ford, Ontario bailed on the entire concept. Meanwhile, federal opposition parties vowed to kill pricing if they came to power.

At the same time, new pipelines bringing diluted bitumen to the east or west coasts, or even to the US, were bogged down in well-funded and -organized opposition by environmental lobbies and First Nations, threatening oil sands expansion plans.

Unable to deliver, would Trudeau continue to receive oil industry support? His unusual step in the summer of 2018 of bailing out a pipeline that hadn't even been built was a sign of how far Trudeau's government would go to support the market.

This book looks at how Canada was persuaded that it could square a circle: We could burn more fossil fuel yet not cook the planet.

CHAPTER 1
BIG OIL TAKES CONTROL

Western society's utter dependence on fossil fuels, and Canada's dependence on imported oil despite the large volumes of black gold the country has produced itself, were both nakedly exposed in the chain of events set off by the fourth Arab-Israeli war in 1973.

Syria and Egypt had launched a surprise attack on Israel to regain territories they lost in the third war in 1967. US President Richard Nixon initiated an airlift to deliver weapons and supplies to Israel, as the Soviet Union was doing for Egypt and Syria. In response, the Arab-dominated Organization of the Petroleum Exporting Countries (OPEC) raised the posted price of oil by 70 per cent. OPEC had been formed in 1960 by four countries in the oil-rich Middle East and Venezuela. By 1973 it comprised eight Middle Eastern nations plus Venezuela, Ecuador, Nigeria and Indonesia. Together they accounted for over half of the world's oil supply.[1] They came together to counter the bargaining power of the Seven Sisters, the world's largest oil companies at the time — Exxon, Mobil, Gulf, Texaco, Chevron, BP and Royal Dutch Shell.

In raising the price, the Arab members used their "oil weapon," imposing an embargo on the US and the Netherlands. They also cut their exports to prevent these countries from acquiring supply elsewhere, while OPEC as a whole further raised the price of oil. Within a year oil increased in price from US$3 to US$12 a barrel.

Canada was originally on the embargo list but within several months was re-classified as neutral, so it was subject only to production and export cutbacks.[2] Nonetheless these measures had a significant impact in Quebec and the Atlantic provinces, which relied on imports for their supplies, raising prices and creating shortages. At the same time, a sharp revision downward in the extent of Canadian reserves raised fears that Canada would not have enough supply for future needs.[3] The minority Liberal government of Pierre Trudeau, 1972–1974, became aware of how vulnerable its foreign oil supplies were.

In contrast, Alberta, Canada's major oil-producing province, experienced a "sudden and massive influx of money that made it the richest province in the country."[4] The province had been the target of American firms that spent large sums on oil and gas exploration, pipelines and refineries ever since the giant Leduc field, about 100 kilometres south of Edmonton, was discovered in 1947. By 1967, American ownership of the Canadian oil and gas sector reached 60 per cent. The Seven Sisters were all active in Canada. Most had taken over Canadian companies to gain access to the country's fossil fuel assets, primarily for exploration and production and only secondarily for distribution and retailing in Canada: They were focused mostly on supplying the American market.[5] After oil was discovered in Alaska, the threat that the Canadian Arctic would fall under foreign domination raised alarm bells when mostly foreign companies began pressuring the Canadian government for access.

Partly in reaction to pressure from the NDP that held the balance of power, the Trudeau government set out to counter the threat by creating a government-owned oil company with the right to explore for oil and gas in remote frontier areas. Such a company would provide a window on the industry so government would have as much information as the oil companies. Given the uncertainty of supply, the government needed to know what resources were available in the territories it controlled.[6] A bill creating state-owned Petro-Canada was passed by Parliament in July 1974 after Trudeau won a majority government. It promised to be "Canadian oil for Canadians."

State ownership in general was not a radical concept in most countries. State-owned oil companies in particular were common during the 1960s and 1970s to ensure local populations received a fair share of oil benefits in the face of the power held by the Seven Sisters. Canadianization of the oil industry was a goal supported by Canadians in every province, including Alberta.[7] Petro-Canada began operating in January 1976.

To head this new entity, Trudeau chose someone who would turn out to have a decisive effect on the politics of oil and climate change over the next four decades — not just in Canada but globally. Maurice Strong was a sort of Janus figure. The attraction for Trudeau, apart from Strong's long-time support of the Liberal Party, was his background in the oil industry. But Strong had another side, too. At the time of the appointment, the forty-seven-year-old Strong had completed his assignment as head of the 1972 United Nations Conference on the Human Environment, and had established the UN Environment Programme in Nairobi, Kenya before returning to Canada. In the long run, it was Strong's environmental face, rather than his energy background, that would matter most, making him alluring to politicians and businessmen with a bigger contradiction to reconcile than Canada being both rich and vulnerable in energy.

Petro-Canada's mandate was to be active on the frontiers — the oil sands, the Arctic and the East Coast offshore areas — and not in the conventional oil and gas sector of Western Canada.

It was provided with $1.5 billion in capital and preferential access to debt financing "as an agent of Her Majesty," reducing the company's costs.[8] It used these funds to purchase the Canadian assets of Atlantic Richfield Canada for $342 million. Two years later it acquired the much larger Pacific Petroleums for over $1.5 billion, tripling the company's oil production with this purchase. In the 1980s Petro-Canada acquired Belgian-controlled Petrofina Canada and British-controlled BP Canada.[9] Within ten years Petro-Canada was a major player on the Canadian energy scene.

PIERRE TRUDEAU AND THE NATIONAL ENERGY PROGRAM

The second oil shock occurred in 1979 after Iranian oil workers went on strike and oil production was slashed. Then the Iranian Revolution drove the Shah and his crony capitalists out of the country, sending prices skyrocketing again, more than doubling in just over a year. Faced with declining oil from Iran, oil companies took dramatic measures to tap other sources. To supply its customers in Quebec and the Maritimes, for instance, Imperial Oil had been importing 100,000 barrels a day from Venezuela. Exxon, Imperial's parent, ordered Imperial to divert 25,000 barrels a day of that to supply its customers in the US. This action threatened to leave some Canadian Imperial Oil customers high and dry. In response, the federal government told Imperial to get the oil back. Other companies importing oil from Iran for Eastern Canada undertook swaps with the US, sending extra Alberta oil southwards in exchange for some American imports to be offloaded for Eastern Canada.[10]

Trudeau lost the 1979 election to Joe Clark and the Progressive Conservatives but came back the following year with a convincing win, campaigning on Canadianization of the energy industry. Oil prices were at an all-time high. Within eight months, Trudeau introduced the National Energy Program (NEP) with broader goals than merely to set up a national oil company. It would increase the federal share of energy revenues, boost Canadian ownership of the industry to 50 per cent of sales revenues and make Canada self-sufficient as an oil producer. The government also introduced a tax to fund Petro-Canada and give grants to Canadian-owned companies to encourage exploration.[11] Economic benefits were transferred to Ottawa and energy users, especially in Ontario and Quebec, at the expense of industry and the producing provinces.[12]

It may have been the most ambitious effort by a Canadian government to intervene in the economy in recent history. It was supported by large Canadian firms that saw an opportunity to expand their oil and gas operations. It was also widely approved by the Canadian public. The Canadian Petroleum Association's own confidential poll suggested that 84 per cent of Canadians supported the move to make the oil and gas industry at least 50 per cent Canadian-owned. This included a plurality in all provinces.[13]

The NEP was not without its flaws. It ran roughshod over Indigenous interests, pre-empting land claims and undermining efforts in regional economic development.[14] But concern about Indigenous rights wasn't what impelled Big Oil to take action. American companies were used to making their own decisions about the kind of access to Canadian energy resources they needed. The NEP put a serious crimp in their plans. The "made-in-Canada" prices for oil and gas allowed Ottawa to discriminate against the US. This raised Big Oil's ire — it moved drilling rigs across the border into the US in front of television cameras and slashed exploration budgets in a version of a capital strike.[15]

Alberta premier Peter Lougheed was so irate that he cut Alberta's crude oil production by 180,000 barrels per day, forcing oil imports to rise and undermining the NEP's self-sufficiency goal. That brought Trudeau to the table, the Parkland Institute's Gordon Laxer notes, and an agreement was soon reached.[16] Albertans were furious by what they saw as a cash grab. "Let the eastern bastards freeze in the dark" was the infamous bumper sticker. Resentment lingered for years even after the Brian Mulroney government dumped the program in 1985.

The NEP was also condemned by the corporate media. *The Globe and Mail*, for instance, published nearly 200 articles and columns that mentioned the NEP in the six months after the legislation was introduced in parliament in October 1980. Most were derogatory, critical or just plain negative. Some headlines proclaimed: "It imposes major economic burdens on all Canadians"; "Energy policy raises spectre of rationing"; "Energy plan will cheat Canadians of oil solvency"; "'Disaster' seen in energy plan"; "Chamber urges protests to stop energy program"; "Liberal policy halts $20 billion energy projects"; "Independent oilmen's president calls NEP a national disaster." As for positive headlines, there were two: "NEP is good for Silverton," because it was a 75-per cent Canadian-owned company; "Self-sufficiency rides on success of NEP." Both of these positive cases occurred toward the end of the six-month sample period, when attitudes had likely already hardened.

BUSINESS COUNCIL ON NATIONAL ISSUES AND THE WESTERN ACCORD
The assault on the NEP was led by the oil industry working in conjunction with a new organization, the Business Council on National Issues (BCNI). Most business organizations represented the interests of particular industries: Canadian Association of Petroleum Producers and its predecessor, the Canadian Petroleum Association, Canadian Gas Association and Canadian Energy Pipeline Association were some important examples in the fossil fuel industry.

A few organizations represented business as a whole and BCNI was the most influential of these. It was formed in 1976 by the CEOs of Imperial Oil and Noranda to protect corporate profits the year after the Trudeau government imposed wage and price controls on the Canadian economy (and the same year Petro-Canada commenced operations). They quickly recruited a blue-ribbon roster of chief executives for the cause.

The BCNI was modelled after the Business Roundtable in the United States. The idea of these groups — they also existed in the United Kingdom, Australia and many other countries — was to develop a rapid response capability by bringing together corporate chief executives who could quickly commit their companies' financial and political resources to the council's goals and actions.[17] The BCNI took on energy as a priority in the early 1980s in reaction to the NEP. However the organization had a serious problem in targeting a specific sector. In representing business as a whole on issues like energy or global warming, the BCNI had to juggle the interests of its members. Alberta oil producers might have different political objectives than Ontario manufacturers, for instance. The oil and gas position prevailed in attacking the NEP and continued to dominate the organization, as befit the importance of energy to the Canadian economy.

The BCNI established its credibility as a player on energy issues by engineering a new accord to replace the NEP. It arranged private meetings with bankers, oil executives and representatives of the Alberta, Ontario and federal governments. On the federal front, the Conservatives made a pact with the premiers and provincial energy ministers not to let energy policy become an issue in the 1984 election campaign that Mulroney won handily.[18]

Trudeau had barely moved out of 24 Sussex Drive when Brian Mulroney, leading Canada's first neoliberal government and working closely with big business and Big Oil, reversed almost all that Trudeau had accomplished in the energy field. Mulroney signalled the energy industry that his government would be sympathetic to its demands by appointing Pat Carney as his energy minister and Harry Near as her chief of staff.

Carney had been an oil industry consultant in the Northwest Territories and a promoter of the Mackenzie Valley pipeline proposal before going into politics. Her mandate as minister was to dismantle the NEP. She was ably aided in this mission by her chief of staff Near, a long-time Conservative Party operative who had been an executive at Imperial Oil for nine years before becoming Carney's top assistant. Within a year of Mulroney's election, the NEP was history when Carney and the energy ministers of Saskatchewan, Alberta and British Columbia signed the so-called Western Accord, which removed any remnant of government intervention and put the Canadian Petroleum Association in charge.

The process through which this was accomplished "marked a striking departure from an open, democratic process of government," observed York University political economist David Langille, who undertook an early study of the BCNI.[19] The Western Accord was a "virtual echo" of BCNI proposals. It deregulated crude oil and natural gas prices, allowing them to rise to international levels. It cut federal taxes and provincial royalties. It based resource taxation on profits rather than revenues, delinking public ownership of the resource and corporate exploitation of it. And it provided exploration and development incentives through the tax system rather than through grants, obscuring these public subsidies to the industry. The net effect was a $1.3 billion annual bonanza for the industry.[20]

The Western Accord was one of Canada's first exercises in neoliberalism, reducing government's ability to intervene in the energy sector, and using its powers to reshape the energy market through deregulation, tax cuts and privatization. This third neoliberal strategy was not used at this time to privatize Petro-Canada. As we've seen, the national energy firm was very popular with Canadians in every province — even Alberta.[21] It would not be privatized until the mid-nineties and another change of government.

With the NEP assigned to the dustbin of history, Near left official politics with a stable of energy industry clients to set up shop as a lobbyist. He later co-founded Earnscliffe Strategy Group, one of Canada's leading lobby firms. Carney was appointed minister for international trade with a mandate to negotiate the Canada-United States Free Trade Agreement, in which energy arrangements between the two countries played a central role. And BCNI became the fossil fuel industry's champion, a role it played over succeeding decades, perfecting its methods of pushing Canadian public policy in directions desired by business. The council changed its name in 2001 to Canadian Council of Chief Executives to better reflect its international interests. (It changed its name again in 2016 to Business Council of Canada, a more benign formulation.)

In his analysis of corporate rule, *The Myth of the Good Corporate Citizen*, journalist and activist Murray Dobbin provided a checklist of the business council's successes in the early years: engineering the Western Accord; forcing through the free trade agreement; defeating Liberal finance minister Allan MacEachern's 1981 budget, which would have closed 165 corporate tax loopholes; removing the teeth from anti-combines legislation, which would have restricted corporate mergers and takeovers (1981); introducing and promoting the goods and services tax (GST), which transferred $18 billion in taxes from corporations to individuals; and convincing the Liberals to abandon their Red Book of enhanced social programs and focus instead on slaying the deficit.[22]

And it has played an important, although largely unacknowledged, role in shaping Canadian and international global warming policy to suit the needs of business. The BCNI's clout was evident in its ability to frame public debate. To find out what the federal government planned for tomorrow, it had been said, look at what the BCNI advocated today. Such an investigation would reveal the concordance between the council's agenda for climate change and Justin Trudeau's eventual promotion of this agenda.

PETER LOUGHEED AND ALBERTA ENERGY OWNERSHIP

Alberta Premier Peter Lougheed would have his moment with the BCNI in the run-up to the fateful 1988 so-called free trade election that established the ground for Big Oil dominance. While Pierre Trudeau was attempting to exercise control over the oil industry to protect Canadian consumers through Petro-Canada and the National Energy Program, Peter Lougheed, a Conservative, was making similar efforts to benefit Albertans.

When Lougheed was elected premier in 1971, government participation in the economy was still common sense in many political circles. He would apply this orientation to ensuring provincial control of resources, something that was written into his genes. But he would do it as a capitalist, following in the footsteps of his grandfather, James Lougheed, one of Western Canada's wealthiest businessmen.

James Lougheed combined law, politics and business into a seamless enterprise. What would be seen later as conflict of interest could also be seen as fusion of interests. When Alberta became a province in 1905, the Dominion government refused to transfer its ownership of natural resources to the new province. James Lougheed worked long and hard to affect the transfer, something that didn't happen until 1930, five years after his death. He was an investor and lawyer at the first oil and gas company in Western Canada, Calgary Petroleum Products, which hit pay dirt in Turner Valley southwest of Calgary in 1914. Revenues resulting from the ownership of the resource went to Ottawa, not Edmonton, where Lougheed was involved in many businesses. Why shouldn't Alberta and its leading businessmen benefit from its resources?

When grandson Peter came to power, it was clear the conventional oil boom that started in Leduc in 1947 was ending due to a lack of new major oil discoveries. If Alberta, and indeed Canada, were to be self-sufficient, oil sands production would need to be ramped up.[23] "The oil sands are owned by the people," grandson Peter declared. "They're not owned by the oil companies." He urged Albertans to think like owners and levy their fair share of royalties, the rent the developer of a resource pays to the owner for its use.[24] He immediately set out to revise the province's royalty structure to increase Alberta's share of revenues and to assert the provincial Crown as owner. He understood that to prosper in the future the province had to maximize returns from its declining non-renewable resource base. The Social Credit government that Lougheed's Conservatives replaced had established a maximum royalty rate of 16.7 per cent and written it into law. Lougheed saw setting the rate so low as "a very serious error in judgment."[25] Lougheed's bid for a larger share of resource revenues raised the hackles of industry because under Social Credit it was used to receiving a bigger slice of the pie.

The "oil sands" are the vast deposits of bituminous sand that underlie much of northeastern Alberta, centred near Fort McMurray. As early as 1906, Canadian and Alberta government agencies started working on ways of separating the sand from the bitumen and turning the bitumen into usable oil. Once the public sector had demonstrated the technical feasibility of the separation process — this occurred in 1950 — five major American-based oil companies took it over. These were Great Canadian Oil

Sands, owned by Sun Oil; and Syncrude, owned by Cities Services; Imperial Oil, Atlantic Richfield and Royalite Oil.[26]

In promoting oil sands development, Lougheed pushed hard for economic benefits for Albertans, including ownership, or equity positions. He set up the Alberta Energy Corporation as a public-private vehicle to directly participate in development of the oil sands and provide Albertans with an opportunity to share in the ownership of provincial resources. (He had the model of the Alberta Gas Trunk Line, a provincial Crown corporation set up by Social Credit Premier Ernest Manning in the 1950s to operate a gas pipeline within the province.[27]) The Alberta Energy Corporation was so successful it would end up decades later as two privatized industry giants: natural gas leader Encana Corporation and oil sands major Cenovus Energy.

Lougheed launched the Alberta Oil Sands Technology and Research Authority to expand oil sands research and development in the province. Its mandate was to develop technologies that could recover oil sands deposits unreachable by surface mining techniques. He created the Alberta Heritage Savings Trust Fund to preserve some of the revenues from non-renewable resources for the benefit of present and future Albertans. According to *Toronto Star* business columnist David Crane, Lougheed was seeking to build a base of companies headquartered in the province that could make decisions in Calgary and Edmonton and not in Houston or Riyadh.[28]

Perhaps Lougheed's most crucial influence on oil sands development was his handling of the massive Syncrude project, which was on the table when he came to power. This project was backed by the Canadian subsidiaries of four major American oil companies. Indeed, oil sands development was largely an American affair. American interests believed that at some point synthetic oil would be necessary and could compete price-wise. Capital for the oil sands came from the US and that was where the markets were.

Lougheed well understood that the first oil sands project, Sun Oil's Great Canadian Oil Sands, an open-pit mining operation north of Fort McMurray on the Athabasca River, which commenced production in 1967, faced challenging financial prospects, at least until the first oil shock, when the price of oil skyrocketed.[29] He was determined that Syncrude succeed. That meant a royalty rate to make the project viable for the private developers, but also one that would return a fair share to the resource owners.[30]

The royalty arrangement hammered out by the parties had some unique features. Instead of setting the royalty at a percentage of production value as it did with Great Canadian Oil Sands, the province would take 50 per cent of Syncrude's net profits. This would work because the federal government granted Syncrude a special exemption from its tax policy, treating the profit-sharing deal as if it was a royalty payment that could be deducted from Syncrude's federal tax bill.[31] Alberta Energy Corporation was also an investor in the project but Lougheed discounted any suggestion that government intervention with private industry was a denial of his free enterprise philosophy. Government participation was simply essential for such a large project.[32]

To complicate matters further, Atlantic Richfield Canada, one of four Syncrude investors, pulled out in 1974 to move its money to the burgeoning oil and gas play at Prudhoe Bay, Alaska. That left a 30 per cent interest on the table. The remaining three companies threatened to pull out unless the federal and provincial governments agreed to provide CAD$600 million in equity. In short order, the governments complied, with $300 million from the federal government for 15 per cent, $200 million from Alberta for 10 per cent and $100 million from Ontario for 5 per cent.[33] The three remaining companies increased their participation to make up the funding. The total investment was CAD$2 billion ($7.5 billion in 2017 dollars). Syncrude began production in 1978. The timing was favourable for the consortium as world oil prices leaped into the stratosphere after the second oil shock. Syncrude and Great Canadian Oil Sands were both profitable during the 1980s, even during the years the NEP was in force.

BRIAN MULRONEY AND THE BCNI BRING FREE TRADE TO CANADA

"Canada is open for business again," Prime Minister Brian Mulroney told a business audience at the Economics Club of New York three months after his landslide election victory in September, 1984. Canada was taking "a fundamental change in economic direction."[34] He had already revised investment rules, making it easier for foreigners to buy Canadian assets or operate their own businesses in Canada, policy changes that would be of interest to American oil companies. After working with energy minister Carney and the Business Council on National Issues to undo the NEP,

Mulroney would work with them once more to ensure an NEP could never again be brought in to deter investment and business. He appointed Carney in her new portfolio as minister for international trade to oversee the negotiations.

The BCNI started looking at trade problems in 1982, when it formed a small task force to investigate how Canadian business could prevent the administration of Republican President Ronald Reagan from building barriers to Canadian access.[35] At a meeting with William Brock, Reagan's trade representative, Brock brought up the idea of comprehensive free trade between the two countries. Canadian CEOs were keen to work on it. But Brock cautioned them that the idea must be seen as coming from the Canadians. Americans proposing free trade would be toxic.[36] It was an idea for the future, though, because little was possible during the Trudeau years, although Peter Lougheed endorsed the BCNI position on free trade and rallied support among Western premiers.

Neoliberalism didn't start in Ottawa the day Brian Mulroney walked in the door; it had already been making its way into official circles. The Royal Commission on the Economic Union and Development Prospects for Canada was appointed by Trudeau in 1982. Chaired by Trudeau's former finance minister, Donald Macdonald, the commission's massive three-volume report, supported by seventy-two research volumes, was presented not to Trudeau but to his successor, Brian Mulroney, in 1985. Among the commission's many recommendations were those presented from a neoliberal perspective — Canada should rely more on market forces and promote free trade with the Americans. It rejected outright the interventionist policies of the recent past.[37]

The Macdonald Commission legitimized Mulroney's abrupt change of direction. It wasn't too long before Mulroney and Reagan met at the so-called Shamrock Summit in Quebec City. They announced that the two countries would begin to explore a possible comprehensive free trade agreement. They made it clear that the idea came from the Canadians, even though it didn't. Negotiations dragged on for several years until an agreement was reached in 1987.

What was unique in the agreement was the inclusion of a separate chapter on trade (and investment) in energy. This chapter is notable for its restrictions on government power. It prohibited the imposition of discriminatory export taxes, something that the NEP used to promote domestic consumption. And it reduced the allowable exceptions to this rule. Any taxes imposed by one country must apply also to its own domestic consumption. Therefore export prices must be the same as domestic prices. Regarding volumes, Canada could restrict exports to the US only in the same proportion that it curtailed domestic production and by no more than the average proportion of Canadian oil exported the US over the previous three years. The NEP's made-in-Canada pricing and energy security for Canadians were out the door — forever, unless a future government somehow found the political support to abrogate the entire agreement.[38]

Peter Lougheed resigned as premier of Alberta in 1985, but he didn't leave politics. He was a staunch supporter of free trade because proportionality meant that a federal government could never again impose an NEP. He may have been thinking of an overreaching federal government, but the reality was that power passed into the hands of big American oil companies. According to Marci McDonald, *Maclean's* Washington bureau chief during the Mulroney years, in her book *Yankee Doodle Dandy*, six large American oil companies lobbied for unrestricted access to Canadian oil. James Baker, Reagan's secretary of the Treasury and a prominent Houston oil lawyer, put the proportionality provisions on the negotiating table.[39] Neither Lougheed nor Carney objected to the measure. In fact they were delighted. "Critics say that the problem with the FTA is that under its terms Canada can never impose another NEP on the country," she said. "The critics are right, that was our objective," Lougheed agreed. "The biggest plus of this [agreement] is that it could preclude a federal government from bringing in a National Energy Program ever again."[40]

The agreement was signed in January 1988, and the federal election held later that year went down in the history books as the free trade election. Liberals and New Democrats were opposed to free trade and together garnered 51 per cent of the popular vote. But the Conservatives won a majority of seats with 43 per cent of the vote and free trade was passed by parliament.[41] A BCNI-sponsored front group, the Canadian Alliance for Trade and Job Opportunities, spent millions of dollars during the campaign promoting free trade. Peter Lougheed and Donald Macdonald co-chaired the group, but it was run by Alcan's CEO, David Culver, who was BCNI chair. Murray Dobbin says the free trade initiative "was the most concerted and massive corporate intervention in any election in the history of the country."[42] Free trade — unfettered investment, really — became a reality when the FTA came into effect on January 1, 1989. It was one of Canada's defining neoliberal moments.

Although Trudeau and Lougheed were opponents politically, both men ultimately failed to achieve their objectives as neoliberal ideology swept political and economic thinking. Market approaches became the order of the day by the time they left active politics, Trudeau in 1984 and Lougheed a year later. While Mulroney undid Trudeau, Lougheed, ironically, would be undone by his own successors at the head of the Alberta PC party.

THE SEVEN SISTERS IN 1988

As we've seen, the 1973 and 1979 oil shocks had profound impacts on the Canadian energy system. They also changed the global industry, but not as much as has been claimed. The shocks were largely a response by oil-producing countries to global domination of prices and supply by the Seven Sisters. That moniker was coined by Italian oil executive Enrico Mattei, head of the Italian state energy company ENI. He was frustrated by the fixed prices and tight control exercised by these companies. Their power rested on their almost exclusive access to low-cost oil in the Middle East, Venezuela and Indonesia. They didn't compete as much as participate in partnerships. This was not new behaviour for Big Oil. To head off declining production during the Great Depression of the 1930s, Shell, Anglo-Persian Oil Company (later BP), Standard Oil of New Jersey (later Exxon) and Gulf Oil agreed to divide up the world's oil reserves and markets. This Depression-era combine also cut prices to drive smaller companies out of business and force those that remained to agree to production quotas.[43]

They emerged from the Second World War more powerful than ever. In 1945, US President Franklin D. Roosevelt met King Abdul Aziz of Saudi Arabia aboard the *USS Quincy* while traversing the Suez Canal. They concluded a secret agreement in which the US would provide Saudi Arabia with military security — military assistance, training and a military base in the country — in exchange for secure access to supplies of oil.[44] Less than a decade later, Iran's democratically elected prime minister, Mohammad Mosaddeq, was overthrown in a coup sponsored by the CIA with British support. Mosaddeq's crime was that he had nationalized the British Anglo-Iranian Oil Company (formerly Anglo-Persian and later BP). His overthrow consolidated the Shah's rule for the next twenty-six years until the 1979 Islamic Revolution and the second oil crisis.[45] Mosaddeq's ouster was aimed at ensuring that the Iranian monarchy would safeguard the West's oil interests in the country. As a result of these and other government interventions, the 1950s and 1960s were golden years for the industry, as it grew rapidly and further consolidated into global giants, a result of explosive demand for gasoline and petrochemicals. The Depression-era combine re-appeared as the Seven Sisters, obtaining near-monopoly control over production and marketing.

As we've seen, to challenge Big Oil, OPEC unilaterally raised oil prices in 1973 resulting in the supply crisis. The big companies subsequently lost much of their direct access to the massive oil resources of the Middle East and other countries. Producers were taking control of their own resources, a pattern attempted in Canada by Pierre Trudeau and Peter Lougheed. In response, Big Oil looked for and successfully acquired oil assets in non-OPEC countries and diversified into coal and metal mining to take up the slack (shedding these activities when the next oil boom occurred).

Thanks to FDR's secret agreement with Abdul Aziz, oil giant Chevron — known earlier as Standard Oil of California — discovered the world's largest oil field in Saudi Arabia after the Second World War. Chevron established the Arabian American Oil Co. (Aramco), using it as a cash cow until 1973, when the Saudi government began buying into the company, completing the takeover in 1980, and later changing the name to Saudi Aramco. In 2018 it was the largest company in the world, with a market value estimated at US$2 trillion, about five times larger than ExxonMobil.

To regain some of the heft it was losing to Saudi Arabia, Chevron acquired Gulf Oil in 1984, nearly doubling in size. As well, Chevron and the other majors looked elsewhere for oil, in the North Sea, Gulf of Mexico, Africa and Asia. Massive oil fields were discovered in the North Sea between the United Kingdom and Norway by the US firm Phillips Petroleum in 1969. Production became economical after the 1973 oil crisis when oil prices quadrupled. Chevron discovered its own North Sea oil field in 1977. By 1978, the North Sea was producing one million barrels of oil a day. No major could afford not to have a North Sea presence.

The pattern repeated itself in the Gulf of Mexico and elsewhere. The Prudhoe Bay oil field in Alaska was operated by BP with partners ExxonMobil and ConocoPhillips. Atlantic Richfield had discovered oil there in the late 1960s, withdrew from its participation in the Syncrude oil sands project to develop its holdings and was eventually taken over by BP.

By the time global warming became a public issue in 1988, the Seven Sisters (Gulf Oil was taken over by Chevron, but France-based Total emerged as a new Sister after a string of acquisitions) had re-established their hegemony over oil politics. With their century-long history of meddling in affairs of state, they would present a formidable obstacle to meaningful action on climate change.

In Canada, the oil sands boom seemed to be over. Suncor was producing 60,000 barrels a day and Syncrude three times that. Together, the two mines were responsible for 16 per cent of total Canadian oil production.[46] They were joined by Imperial Oil's Cold Lake Mine in 1985. With oil prices hovering in the mid-teens, no other projects were in the pipeline. But American companies had won guaranteed access to Canadian energy resources. The stage was set for an oil sands boom whenever oil prices took off again.

CHAPTER 2
CONTAINING ENVIRONMENTALISM

The modern environmental movement and the "endless war" waged by polluting industry to resist it both got under way in 1962.[1] Rachel Carson's game-changing book, *Silent Spring*, raised the alarm that pesticides were poisoning wildlife and endangering human health. She cautioned that bald eagles eating fish laden with DDT were threatened with extinction. Her conclusion was clear: Industrial practices and products had to be changed. But even before her book was released, Bruce Harrison, then-manager of environmental information for the Manufacturing Chemists Association, launched an all-out attack on her credibility.[2] Working with PR executives from Shell, DuPont, Dow and Monsanto, Harrison trained his guns on her, using the emerging practice of "crisis management," which has been described as a melange of "emotional appeals, scientific misinformation, front groups, extensive mailings to the media and opinion leaders and the recruitment of doctors and scientists as 'objective' third party defenders of agrichemicals."[3] He would use the same techniques thirty years later to attack the findings of the Intergovernmental Panel on Climate Change, the key global body on the issue.

The war had to be endless because energy, forestry, mining and chemical companies would always be polluting the environment and destroying habitat and citizens would always be trying to stop them.

The battle to deny business's environmental impacts had an unsettling parallel, extending over many decades, in the tobacco industry's efforts to deny the health effects of its products. It was the tobacco challenge that inspired PR firms to pioneer anti-public-interest propaganda during the 1950s after the first studies suggesting a link between smoking and lung cancer were published. That campaign was orchestrated by John Hill, founder of Hill+Knowlton, today one of the largest PR firms in the world. This work is called "grassroots propaganda" because it is directed at the public. Its purpose is to turn public opinion against citizen activism and portray industry in a good light. Cancer researcher Devra Davis in her 2007 book, *The Secret History of the War on Cancer,* sums up John Hill's brilliant strategy, still in use today in many industries:

> [C]reate doubt. Be prepared to buy the best expertise available
> to insist that more research is needed before conclusions can
> be reached. Whenever new studies emerged on the hazards
> of smoking, the tobacco industry would flood reporters'
> in-boxes with counterarguments asserting that nothing had
> been proven. It would marshal its own experts to magnify the
> appearance of a scientific debate long after the science was in
> fact unequivocal. [4]

Treetops propaganda, which uses similar techniques to influence decision-makers, would come later, in the form of neoliberal think tanks. Both types of propaganda dissemination would be needed to protect the fossil fuel industry's long-range growth plans. But despite the efforts of the PR industry to discredit Carson and others, public concern about the environment was widespread and growing. The Environmental Defense Fund was created in 1967 to "sue the bastards" by going to court to successfully stop the Suffolk County Mosquito Control Commission from spraying DDT on the marshes of Long Island. DDT was subsequently banned in the United States and Canada in the early 1970s and in most developed countries by the 1980s. The first Earth Day was held on April 22, 1970 as a national teach-in about pollution, resource depletion and waste produced by a consumer-oriented society. Twenty million people participated in peaceful demonstrations across the United States. The same year the Richard Nixon administration created the Environmental Protection Agency and Congress passed the *Clean Air Act*, followed two years later by the *Clean Water Act*, two of the most stringent environmental protection actions ever attempted by the American government. In Canada, the Trudeau government passed its clean air act and established the Department of the Environment in 1971 in response to public pressure.

SPACESHIP EARTH AND THE LIMITS TO GROWTH

The environmental movement was buttressed by a small band of scientists and economists concerned that two decades of explosive growth had led to the rapid expansion of the human population and the drawing down of the planet's resources. Kenneth Boulding (1910–1993) was an American Quaker and committed pacifist. His 1966 essay, "The Economics of the Coming Spaceship Earth," challenged prevailing economic ideas about an open system with unlimited resources and an unlimited capacity to absorb waste. He labelled this dominant view the "cowboy" economy, evoking images of the endless American frontier. By the postwar period the frontier had been completely colonized. You could no longer amble over the next mountain range, expect to take all the water for a gold-mining operation and not worry about toxic run-off. Humanity had expanded the scale of its activities to such a point that pollution and resource extraction threatened the long-term survival of society. As a result, Boulding argued, we need to imagine a "spaceman" economy where the earth is a spaceship with limited resources and limited capacity to absorb pollution.[5] Spaceship earth was a metaphor the public could understand and get behind, coming as it did in the midst of the space race between the United States and the Soviet Union.

Romanian-born mathematician and economist Nicholas Georgescu-Roegen (1906–1994) had a more direct impact on the development of a new social science discipline called ecological economics. He used the laws of thermodynamics to demonstrate that traditional economics failed to account for the wastes resulting from production.[6] Pollution was not external to the economy, as most economists would have it, he argued, but intrinsic to it. His work challenged fundamental assumptions of mainstream economics, including the dogma that there are no limits to the growth and expansion of market economies.[7] Georgescu-Roegen was recognized as a significant contributor to the field of economics, but never received the Nobel Prize, reflecting, perhaps, the bias toward pro-growth views. His work was largely ignored or else attacked by prominent pro-growth mainstream economists like Robert Solow and Joseph Stiglitz, both of whom did win the Nobel Prize.[8]

Herman Daly was Georgescu-Roegen's most famous student, developing an alternative to the conventional growth-is-good paradigm, which he labeled the "steady-state economy" in his 1977 book of the same name.[9] The economy must conform to global environmental constraints, he maintained. Growth beyond global limits will produce more social and economic costs than benefits. A steady-state economy would support qualitative development but not quantitative growth, a view, like Georgescu-Roegen's, that was virtually ignored by the mainstream economic consensus.

German economist Fritz Schumacher, who was an adviser to the British government on its coal industry, was probably the first author to use the term "natural capital." In his 1973 classic, *Small is Beautiful*, Schumacher argued that Western economies were making a grievous error in failing to distinguish between the income they derived from nature and the capital it provided. As a result they were using up nature's capital "at an alarming rate."[10] They must become concerned about conservation and try to minimize the current rate of drawing down the stock of resources.

Several years later, biologists Paul and Anne Ehrlich, in their famous book, *Extinction: The Causes and Consequences of the Disappearance of Species*,[11] attempted to demonstrate how the disappearance of biodiversity directly affects ecosystem functions that underpin human well-being, thus justifying improved protection of biodiversity.[12]

Spaceship earth, steady-state economy, limits to growth and ecosystem functions were heretical ideas in mainstream economic thinking, but spoke to the concerns of scientists and others grappling with ways to highlight the looming environmental crisis facing humanity. But for industry worse was yet to come. In 1968, thirty-six prominent scientists, economists and business leaders established the Club of Rome to gain a holistic understanding of problems facing the world and what their solutions might look like.

With funding from the Volkswagen Foundation, the group asked computer experts at the Massachusetts Institute of Technology what would happen if humanity continued to consume resources at then-current levels. The result was *The Limits to Growth*, published in 1972. It eventually sold sixteen million copies in thirty languages.[13] Despite obvious limitations in the quality and quantity of the data and computer modelling capacity available at the time, the study delivered an unambiguous answer: If humanity did not moderate its resource use, society and the environment within which it lives would likely collapse by 2100.

ENTER MAURICE STRONG

But what if instead of the environment being destroyed by growth, it could be claimed that the environment would be saved by growth? Or at least the case could be made that growth is necessary to protect the environment. This framing of the environmental problem would soon adopt the name "sustainable development" and provide a perfect counter to fears of environmental collapse. This was where Maurice Strong came in. Before the idea of a Petro-Canada headed by Strong came across Pierre Trudeau's desk, Strong had taken on the task of making the case for environment *and* development as secretary-general of the 1972 UN Conference on the Human Environment, held in Stockholm three months after *The Limits to Growth* was published. In his opening speech of the conference on June 5, 1972, Strong declared: "There is no fundamental conflict between development and the environment." And he applauded "the new synthesis that is now emerging."[14]

The conference had not started out this way. From its very beginning, industrial pollution was an international problem that didn't respect borders. Just ask Rachel Carson. Acid rain was already a problem in the mid-1960s. Sulphur dioxide emissions from power plants in northern Germany were affecting surrounding countries in Scandinavia just as they were doing in Canada, the US and elsewhere. In 1968, the Swedish government proposed to the UN Economic and Social Council the idea of having a UN conference on human interaction with the environment. The UN General Assembly gave the conference the go-ahead a year later. This was the first UN conference to focus on one topic. It became a preferred way of dealing with international problems for the global community. UN secretary-general U Thant agreed to Swedish diplomat Sverker Astrom's suggestion that Strong be appointed to lead the conference.[15]

Why was Maurice Strong selected? When he was given this assignment, he had several decades of experience in the oil industry. He worked for Calgary-based oil and gas company Dome Petroleum in the 1950s. He developed service station sites for Chevron in East Africa. He then took tiny Canadian-based Ajax Petroleum and turned it into a much larger Norcen Energy Resources, which was later bought by US-based Union Pacific Resources in the 1990s. His success in oil and gas led to his being hired as president of Power Corporation, which was then a family holding company soon to be sold to Quebec's powerful and politically connected Desmarais family. Strong maintained an impeccable connection to the federal Liberal Party, at one point hiring future Prime Minister Paul Martin, Jr. as his executive assistant. Strong's success at Power led to his appointment by Lester Pearson as the first president of the Canadian International Development Agency, which provided aid to developing countries. That appointment led to close relations with leaders of the developing world.[16]

And that chain of connections might answer the question as to why an oil executive would be appointed to lead an environmental conference, especially an oil executive who seemed never to have shown interest in the environment "except as something to be used," as Toronto journalist Elaine Dewar put it in her book *Cloak of Green*.[17] Strong told Dewar he was selected by UN officials because they saw him as having influence in developing countries such as Brazil, which led the initial opposition to the conference. The sponsoring Swedish government believed Strong could bring the Brazilians on board. (He certainly did that in 1992, when he led the Earth Summit held in that country.) Like Brazil, many developing countries believed that the conference would lead to extra restrictions being placed by rich countries on trade and development with them. Strong was able to overcome some resistance because of his reputation gained through his work at the Canadian International Development Agency and by marshalling financial support from the World Bank, the financial institution closest to the developing world.

Bringing developing countries on board was certainly one reason for his appointment to head the Stockholm conference. Another was that by the late 1960s, Strong had forged close links to the Rockefeller Foundation and David Rockefeller, the power behind the Trilateral Commission.

For generations of progressives, the Trilateral Commission was the organization that ran the world. It was established the year after the Stockholm conference and the year of the first oil shock by Rockefeller, head of the family-controlled Chase Manhattan Bank, and Columbia University professor Zbigniew Brzezinski, who became Jimmy Carter's national security adviser. Their goal was to create an international forum to foster closer co-operation among the core industrialized democracies of Europe, North America and Japan. It included 250 members of the financial, industrial, political, bureaucratic and media elites, carefully selected and screened by Rockefeller.[18]

When Rockefeller set up the commission, the business priority was to guide national economies toward an international marketplace — globalization, in other words. Such a reorientation of economic priorities would require dramatic changes in governmental policies in most industrialized nations. But before this could happen, one problem had to be addressed. An "excess of democracy" was the subject of the commission's first book.[19] As the book's authors saw it, "special interests" such as environmentalists, social justice advocates and equal rights organizations; an overemphasis on social welfare programs; a "bloated" bureaucracy and too much emphasis on protection for workers and national economies all combined to bog government down in ineffectuality. The solution the book offered was that government must become stronger to resist the demands of citizens, except for those citizens who were business executives. In this case, government must become weaker. The pacification of radical environmentalism would bring great relief to Trilateralists.

Strong met Rockefeller in New York in 1947, when he applied for a job at the fledgling United Nations organization. Rockefeller had charge of the UN account at the family's Chase Bank. They developed a close relationship over the years. The Rockefeller Foundation provided a grant to Strong to run the Stockholm conference office and appointed him a trustee of the foundation between 1971 and 1977, the years during which he led the conference and then was executive director of the UN Environment

Programme. These were posts from which he could promote the interests of Rockefeller and his fellow global power brokers. Strong became a member of the Trilateral Commission in 1976.[20]

SUSTAINABLE DEVELOPMENT

Strong's Stockholm conference undermined threats to global business by tying environmental protection to economic The conference produced a declaration of twenty-six principles. They covered the spectrum from the environment to the economy. On the environment side, they declared that natural resources must be safeguarded (#2) and the earth's capacity to produce renewable resources must be maintained (#3). But on the economy side, they state that development is needed to protect the environment (#8) and that environment policy must not hamper development (#11).[21] These principles are irreconcilable.

Nonetheless, environmental protection and economic development became increasingly twined over the next fifteen years. The first document that specifically argued for sustainable development was *The World Conservation Strategy: Living Resource Conservation for Sustainable Development*. This report was produced by three moderate non-governmental organizations, and Strong worked with all of them. Strong chaired the lead group, the International Union for Conservation of Nature. Co-authors were the World Wildlife Fund International (Strong was a vice-president from 1978 to 1981) and the United Nations Environment Programme (Strong had been its first executive director from 1973 to 1975). The report identified the main agents of habitat destruction as poverty, population pressure, social inequity and the lack of free trade.[22] While some of these are undoubtedly correct, one would be hard-pressed to demonstrate that free trade protects the environment. Strong and his co-authors didn't have the benefit of thirty-five years of experience with the destructive impact of free trade on the environment to temper their conclusion. And somehow capitalist exploitation of nature had conveniently disappeared. "Conservation for Development" is a phrase worthy of George Orwell's *1984*.

But it was the 1987 Brundtland Commission chaired by Norwegian prime minister Gro Harlem Brundtland and included Maurice Strong that put sustainable development on the map. The commission report, *Our Common*

Future, a wildly popular document, claimed that we were unable to protect the environment unless we had enough economic growth to lift the world's developing nations out of poverty. In his 2000 book, *Where On Earth Are We Going?* Strong called this "eco-development," but Brundtland's "sustainable development" is what caught on and became the framing for the UN Conference on Environment and Development held in Rio de Janeiro in 1992, justifying the muted response to the threat of global warming.[23]

Brundtland tied pollution to the idea of social progress. To deal with over-consumption by wealthy nations and bone-wrenching poverty in poor ones, Brundtland defined sustainable development as development that "meets the needs of the present without compromising the ability of future generations to meet their own needs."[24] But this noble-sounding principle would be achieved by a vast scale-up in industrial development. This seemingly preposterous proposal would be achieved by greater private investment in the Third World and by the development of new technology, which would make production more efficient and reduce pollution.

Sustainable development held appeal for rich and poor nations alike: Both could continue to expand economic activity.[25] By this time, *The Limits to Growth* was a relic. In the 1970s, the ecological crisis foretold catastrophe and disruption. By the 1980s, the ecological crisis had become a challenge for business and an opportunity to open new markets. We didn't need to change society, as Rachel Carson argued. All we needed was the right mix of innovative technologies and institutional arrangements, and business would take care of any environmental problems.[26] Capitalism created the environmental crisis, capitalism could solve it, Brundtland reassured us.

THE RISE OF NEOLIBERALISM

The direct assaults by the public relations industry and the rhetoric of sustainable development had some success in taming radical environmentalism, but more would be needed. A third front was launched by neoliberal economists. A network of neoliberal think tanks was established around the world in the 1970s and 1980s to promote the dominance of the market and to attack the legitimacy of governments to intervene in the economy.

If any single individual can be said to be the originator of neoliberalism, it is Austrian economist and philosopher Friedrich Hayek (1899–1992).

He could have been anticipating Pierre Trudeau and the National Energy Program when he wrote in his famous 1944 polemic, *The Road to Serfdom*, that "[planning] would make the very men who are most anxious to plan society the most dangerous if they were allowed to do so . . . From the saintly and single-minded idealist to the fanatic is often but a step." The planner and co-ordinator, Hayek opined, was little more than an "omniscient dictator."[27] Only the market can make the right decisions.

Three years after *The Road to Serfdom* was published, Hayek invited leading European and American intellectuals of various free-market persuasions — Milton Friedman, Lionel Robbins and Karl Popper among them — to meet at the Hôtel du Parc in Mont Pèlerin, a village overlooking Lake Geneva in Switzerland. He asked them to consider the rise of postwar demands for social and economic rights and the creation of the welfare state. He also charged them to develop theories and strategies to counter the Keynesian orthodoxy that dominated the intellectual landscape.

Hayek was preoccupied with the world created by his long-time adversary, John Maynard Keynes. He was particularly troubled by Keynes's prescription for government intervention in the economy to achieve full employment and assure a sufficient level of demand. Following the Keynes approach would, Hayek argued in *The Road to Serfdom*, lead to a nightmare world of collectivism and socialism, to slavery, not freedom. When Hayek invited his free-market colleagues to the lakeside spa, they were a beleaguered minority, outgunned by Keynesians, socialists and Marxists. During their ten days of talk they agreed to form the Mont Pèlerin Society, with a mandate to work toward a market state — an individualistic, non-egalitarian society, governed by market transactions.

They believed, however, that the market society they desired would not come about without concerted political effort and organization. They were radicals, not conservatives, who came to demand dramatic government action to create and enforce markets. As Marxist geographer David Harvey explains in *A Brief History of Neoliberalism*, for neoliberals "the role of the state is to create and preserve" strong property rights, free markets and free trade, using force if necessary, to guarantee the proper functioning of markets. "If markets do not exist (in areas such as land, water, education, health care, social security, or environmental pollution) then they must be created, by state action if necessary."[28]

Hayek and his Mont Pèlerin Society colleagues understood government would have to change if they were to successfully reverse Keynesianism and halt government intervention in the market. But in postwar Europe and North America, they couldn't hope to capture and reorganize political power by entering politics directly, given the hostile climate of ideas and the dominance of the Keynesian consensus. "To capture political power, they would first have to alter the intellectual climate," writes Timothy Mitchell, a political theorist at Columbia University.[29] And to alter the intellectual climate, Hayek wrote in an influential 1949 essay, "The Intellectuals and Socialism," neoliberals would have to influence what he termed "professional secondhand dealers in ideas" who control the distribution of expert knowledge to the public. They are "journalists, teachers, ministers, lecturers, publicists, radio commentators, writers of fiction, cartoonists and artists," and even scientists and doctors.[30] They "may be masters of the technique of conveying ideas but are usually amateurs so far as the substance of what they convey is concerned," he wrote.[31] Nonetheless, they are the people who, more than any others," decide what views and opinions are to reach us, which facts are important enough to be told to us and in what form and from what angle they are to be presented."[32] Hayek had outlined a grand scheme of propaganda dissemination.

To accomplish the further task of influencing second-hand dealers in ideas, Hayek understood that he and his colleagues would need their own network of "dealerships," or think tanks. "Backed with funds from corporations and their owners, usually channeled through private foundations," Mitchell writes, "think tanks repackaged neoliberal doctrines in forms that 'second-hand dealers' could retail among the general public. Doctrine was supported with evidence presented as 'research'" which was then packaged into books, reports, studies, teaching materials and news stories and distributed to news organizations and other second-hand dealers.[33] The Mont Pèlerin Society became the central co-ordinating body for what was to become the neoliberal counter-revolution.

Hayek recruited British businessman Antony Fisher to put his plan into operation. Fisher made a fortune by introducing factory-farmed chicken in Britain after the Second World War. He had read a condensed version of

The Road to Serfdom in *Reader's Digest,* and it had a profound impact on him. He sought out Hayek, then teaching at the London School of Economics. Hayek "warned Americans and Britons about the dangers of expecting government to provide a way out of the economic danger," Fisher recounted in his own book, *Must History Repeat Itself?* He wrote that ". . . the decisive influence in the great battles of ideas and policy was wielded by the intellectuals" whom, following Hayek, he characterized as the "second-hand dealers in ideas."[34] Fisher believed that capitalism produces more wealth and distributes it more fairly than any amount of government intervention ever could.

Fisher wanted to go into politics to defeat the Labour government of the day, but Hayek advised him not to waste his time. Instead, he should become involved in the "great battle of ideas" by creating "a scholarly research organisation to supply intellectuals in universities, schools, journalism and broadcasting with authoritative studies of the economic theory of markets and its application to practical affairs."[35] Fisher became lifelong friends with Hayek and, in 1954, was invited to join the Mont Pèlerin Society. The following year Fisher established the Institute of Economic Affairs (IEA) in London. This would be the prototype for the hundreds of neoliberal think tanks that followed.

"It was ten years before the IEA could truthfully claim to be exerting an appreciable influence on debate, and at least another five before its success looked beyond doubt," Fisher's biographer reports.[36] The organization's preferred product was a lively ten-thousand-word monograph aimed at journalists writing for quality newspapers, urging them to review or write about the publication. Each monograph was accompanied with an executive summary and news release for busy readers. Make no concessions to existing political or economic realities, but stick to neoliberal doctrine, was the guiding principle of all IEA publications.[37]

Fisher provided a model that conservative and libertarian business executives could use to turn public and elite opinion away from support for interventionist governments and back to a favourable view of capitalism and the market. Their collective impact was instrumental in shifting the British climate of ideas to such an extent that Margaret Thatcher could sweep into office in 1979 with an agenda to bring "the British economy under the discipline of market forces."[38] One supporter declared "had it not been for the Institute of Economic Affairs, there would have been no Thatcher Revolution. They prepared the ground."[39] Not only did the Fisher model colonize Britain, it was exported to the United States, Canada, Australia and further afield.

THE FRASER INSTITUTE

Global warming was more than a decade away when the Fraser Institute was launched in 1974. It was a period of deep anxiety for British Columbia business leaders. They were dismayed by the actions of the Pierre Trudeau government in Ottawa, which was aggressively promoting participatory democracy and a just society through worrisome policies like universal health care, regional development, job creation, environmental protection, expanded social programs, independence from US influence and, very soon, energy independence. They were even more alarmed by Dave Barrett's New Democratic government in Victoria, elected in 1972, which had implemented many controversial measures such as the *Land Commission Act*, which froze the conversion of agricultural land to urban uses, and the *Residential Tenancy Act*, which imposed rent controls and other constraints on the unfettered rights of property owners.[40]

Some business executives began hatching plots to topple Barrett. *Barron's* magazine referred to B.C. as the "Chile of the north," calling to mind clandestine interventions by the Central Intelligence Agency and the American government into that country a year earlier. But the propaganda approach was soon settled on. Noranda Mines chair Alf Powis, who helped organize the Business Council on National Issues, felt "what was needed was a think tank that would re-establish the dominance of free enterprise ideas, the values of the market, and property rights."[41]

Antony Fisher was invited to Vancouver by economist Michael Walker, who knew of Fisher's work through Walker's lifelong friend, Milton Friedman, and forestry executive Patrick Boyle. They asked Fisher to replicate the success of the Institute of Economic Affairs, and followed Fisher's formula for funding, projects, experts and dissemination:[42] Employ a core group of researchers and contract with sympathetic academics to conduct specific studies that support neoliberal doctrine; disseminate the results through conferences, books, reports and news and opinion pieces to reach relevant second-hand dealers in ideas in academia and the media; finance the enterprise by appealing to wealthy conservative business executives and their foundations and to corporations that can benefit from institute activities.[43] At first the new institute was greeted with a lukewarm response from the business press. But within a few years, it achieved success on all fronts, lining up corporate and foundation supporters, academic collaborators and media cheerleaders.

Fisher established two more think tanks, the Manhattan Institute for Policy Research with William Casey, a Wall Street speculator and later Ronald Reagan's CIA director, and the Pacific Research Institute, with local business leader James North in San Francisco. As with the Fraser Institute, they chose geographical names rather than trumpeting their ideological purpose. Fisher received requests from business executives around the world to help them set up similar organizations in their own countries to promote the free market and individual liberty. In 1981, he established the Atlas Economic Research Foundation (later, the Atlas Network) to provide a turnkey system for creating think tanks. Fisher set up Atlas in San Francisco, where he was living at the time. After his death, Atlas moved to Arlington, Virginia to be near the nation's lawmakers. The organization is named after the Greek Titan, Atlas, who was condemned to hold up the sky for eternity. It was inspired by a quotation from Greek mathematician and scientist, Archimedes, who said, "Give me a lever and I will move the world." The lever was neoliberal propaganda and the world has certainly been moved.[44]

Atlas set up dozens of these organizations in the US and around the world. Earlier American think tanks like the Heritage Foundation and Cato Institute — both inspired by and connected to the Institute of Economic Affairs — had targeted policy-makers in the nation's capital. The new crop of think tanks turned their guns on state and local governments. They needed to change local opinion about the economy and environmental regulation, where states and cities play a large role.

By the end of the 1980s the country was saturated with business-backed think tanks, many with a connection to central command at the Mont Pèlerin Society. Twenty-two of seventy-six Ronald Reagan economic advisers in 1980 were members.[45]

THE GREENHOUSE EFFECT

While big business was put on the defensive in the early 1960s by Rachel Carson's critique of chemical pollutants, an even more profound threat to business-as-usual was coming to light in scientific circles. The greenhouse effect was well-known during the 1960s. Scientists talked about the impact of carbon pollution leading to an increase of two, four or even six degrees of warming. American geochemist Dave Keeling of the Scripps Institution of Oceanography first began measuring the amount of carbon dioxide in the atmosphere in the late 1950s from his station at Mauna Loa Observatory in Hawaii. He found carbon dioxide pollution to be growing at an annual rate of 0.7 parts per million, starting at 313 ppm in 1958. (The rate would escalate to 2.1 ppm after the year 2000.)[46] A 1965 report by a panel of US President Lyndon Johnson's Science Advisory Committee warned that humankind was "unwittingly conducting a vast geophysical experiment." It was injecting carbon dioxide into the atmosphere at such a rate that by the year 2000 the concentration of atmospheric carbon dioxide would increase by 25 per cent, with "measurable and perhaps marked changes in climate," changes that would be deleterious to humans.[47] The panel was off the mark by underestimating the vast scale-up in energy use. By 2010, atmospheric carbon dioxide was up 40 per cent to about 370 ppm since the start of the industrial age.

The relationship between levels of carbon dioxide and temperature was a subject of intense research and speculation among scientists. At a Canadian climate seminar sponsored by the Department of the Environment in Regina in 1981, meteorologist Kenneth Hare, chair of the Climate Program Planning Board that studied ocean climate modelling, presented evidence that within fifty years a doubling of atmospheric carbon dioxide could produce a rise of two degrees in average temperatures around the globe.[48] Hare didn't say this would lead to dramatic change, but a 1983 US Environmental Protection Agency report did. It predicted a two-degree increase by the middle of the twenty-first century and concluded "changes by the end of the 21st century could be catastrophic taken in the context of today's world. A soberness and sense of urgency underlie our response to a greenhouse warming."[49]

SHELL KNEW

It's not like Big Oil companies like Exxon and Royal Dutch Shell didn't know about the increasing levels of carbon dioxide in the atmosphere due to their activities. "The idea of global warming was first highlighted in a Shell report on its coal sector in 1979," writes Keetie Sluyterman, a specialist in Dutch business history at Utrecht University, in a study of Shell's strategies for dealing with environmental issues.[50] The "greenhouse effect," was acknowledged in the report but Shell did not plan to "undertake any independent initiative." Instead it would "support highly qualified organisations that might act on behalf of suppliers and consumers." Nothing happened. (Shell later sold its coal unit to Anglo American plc.) An internal review of health, safety and the environment in 1985 mentioned the greenhouse effect again and noted that a working group would continue to monitor the situation and brief Shell's operating companies. "Thus, industry was already thinking about the problem before it became a major issue in the media," Sluyterman noted.[51]

Thinking but not acting. The strongest indication of Shell's awareness of the consequences of burning fossil fuels was a twenty-eight-minute film it produced in 1991, *Climate of Concern*. The film was intended for viewing in schools and universities. It warned of "extreme weather, floods, famines and climate refugees as fossil fuel burning warmed the world," reported *Guardian* columnist Damian Carrington and *De Correspondent* journalist Jelmer Mommers, who saw the film.[52] "Global warming is not yet certain," the film says, "but many think that to wait for final proof would be irresponsible. Action now is seen as the only safe insurance."

Shell did take action, but not the kind that would reduce the risk of global warming. After the film was released — and withdrawn — Shell joined several new organizations established to divert attention away from the fossil fuel causes of global warming and onto the now convenient notion of sustainable development — the Business Council for Sustainable Development (later headed by Shell's chair) and the working party on sustainable development for the International Chamber of Commerce (headed by an official from BP). Meanwhile, Shell continued to invest tens of billions of dollars in oil exploration in Alaska, Alberta's oil sands, Brazil's and Mexico's offshore reserves and gas and oil fracking wherever Shell could get a piece of the action.

THEY ALL KNEW

The pattern repeated itself at each of the Seven Sisters: They knew, they carried on as if they didn't know and later they denied it was happening. Exxon knew about climate change as early as Shell. A 1980 report by Imperial Oil, Exxon's Canadian subsidiary, admitted that "there is no doubt that increases in fossil fuel usage and decreases of forest cover are aggravating the potential problem of increased carbon dioxide in the atmosphere." The report was uncovered by the watchdog organization the Narwhal at the Glenbow Museum in Calgary.[53] It had been distributed from Imperial Oil headquarters in Toronto to Exxon managers in the US and Europe. The statement was ironic, coming from a partner in the Syncrude oil sands project that opened for business a year earlier, increasing fossil fuel usage while clear-cutting vast swaths of forest cover to prepare the site.

A 1980 Exxon report laid out the company's plans to deal with the growing problem. "[Carbon Dioxide] effect: Exxon-supported work is already underway to help define the seriousness of this problem. Such information is needed to assess the implications for future fossil fuel use. Government funding will be sought to expand the use of Exxon tankers in determining the capacity of the ocean to store carbon dioxide."[54] The Narwhal's research indicates that Imperial Oil and parent Exxon engineers and scientists were aware of the impact of air pollutants — including carbon dioxide — as early as 1970.

A 1978 presentation by a top technical expert at Exxon's research and engineering division warned Exxon scientists and managers that a doubling of carbon dioxide concentrations in the atmosphere would increase average global temperatures by two to three degrees Celsius and as much as ten degrees Celsius at the poles.[55] Quick action was needed, he concluded. Exxon responded swiftly, setting up an extensive research program to sample carbon dioxide emissions and develop climate models.

The company would spend a decade on this research, becoming one of the most knowledgeable organizations in the world about global warming. It needed, after all, to have a comprehensive understanding of the potential risks to its business. But toward the end of the 1980s, funding for the research stopped and Exxon began putting its money into climate change denial.

Perhaps it realized that the risks were so great the company might not survive. But other factors were at play as well. The collapse of oil prices in 1986 as a result of an oil glut led to Exxon laying off many employees, including those working on climate. As well, CEO Lawrence Rawl, who took over company leadership in 1985, pushed the organization to central-ize its strategy process, leaving little room for Exxon scientists to consider alternate scenarios. One was the possibility that society's concerns would lead to the risk of increased regulation and diminished reputation, both of which would flow through to the bottom line.[56] Rising fear of stringent regulation of the company's activities was a factor in its decision to shut down carbon dioxide research and promote denial.

While Exxon was doing its climate research, the American Petroleum Institute ran a task force between 1979 and 1983 to monitor and share information developed by Exxon and the other members.[57] These included Mobil, Amoco, Phillips, Shell, Texaco, Sunoco, Sohio (owned by BP), Standard Oil of California and Gulf Oil, the latter two predecessors of Chevron. Some of these corporations ran their own research units because of their concerns about risk to their bottom line, but none were on the scale of Exxon's operation. In 1983, the American Petroleum Institute decided that influencing government — lobbying — was more important than sharing information. The environmental unit was transferred to the political department to promote the political and economic interests of the members.

Once again Big Oil was following the playbook of the tobacco industry. When scientific studies linking smoking to lung cancer began appearing in the 1950s, the industry's response was to set up the Tobacco Industry Research Council, with science and public relations mandates. Money was no object, one advertising executive proclaimed, since the industry "had more money than God."[58] The purpose of the science campaign was to sup-port research that countered work that was finding links between smoking and disease. Using Hill+Knowlton's resources, tobacco advertising research-er Richard Pollay reports, "the essential tactic was to gather, reproduce, and scatter seeds of scientific doubt to magnify and maintain the appearance of a scientific controversy. The world was scoured for scientific and medical opinions that contradicted or confounded the ever growing evidence of tobacco's carcinogenic nature."[59] The PR effort was based on this research. By tying its environmental research to political and PR objectives, Big Oil was following this model.

JAMES HANSEN AND THE INTERGOVERNMENTAL PANEL ON CLIMATE CHANGE

Two events kicked off global action on climate change in 1988. That they occurred in the same year is coincidental. But they were crucial to the global effort to combat climate change.

In June, climate scientist James Hansen of the National Aeronautics and Space Administration told a US Senate committee that it was 99 per cent certain that the warming trend of recent years was not a natural variation but was caused by a buildup of carbon dioxide and other gases in the atmosphere. "It's time to stop waffling so much and say that the evidence is pretty strong that the greenhouse effect is here," said Hansen, the director of NASA's Goddard Institute of Space Studies in New York City.[60] The story made the headlines of virtually every major newspaper and was carried on many network news shows, perhaps because the US was experiencing a scorching summer. It was one of the hottest on record to that point and culminated in a prolonged drought.[61] There was also Hurricane Gilbert that struck the Caribbean and caused over US$1 billion in damage, and a chunk of ice one hundred miles long and twenty-five miles wide broke off from Antarctica. The crowding together of these dramatic events into a short period heightened public concern about climate and weather.

The second event had less public impact but was more salient. It was the launch later in the year of the Intergovernmental Panel on Climate Change (IPCC).[62] This organization was created by two UN affiliates, the World Meteorological Organization and the United Nations Environment Programme. Its goals were broad and ambitious: to assess available scientific information on climate change and to formulate realistic response strategies. The initiative brought together national governments from around the world and put them in the same room — figuratively speaking — with an exceptionally large number of qualified experts in an assessment process involving multiple worldwide peer reviews. This awkwardly structured body of scientists and politicians would become the reference point for debates on climate change.

Although the World Meteorological Organization and United Nations Environment Programme were the official sponsors, the US was key in creating the IPCC. As the largest financial contributor to the UN system, the US has outsized influence on both of these organizations. And it had good reason to exercise its influence. The US was the biggest source of greenhouse gas emissions and any abatement measures could have severe impacts on the economy. Powerful fossil fuel lobbies — together with a compliant Republican White House — were opposed to any kind of action on climate change.

Nor could key government agencies agree on what to do. They were all over the map. The National Research Council's assessment emphasized uncertainties and advocated a cautious wait-and-see approach. The Environmental Protection Agency, in contrast, spoke of catastrophic consequences resulting from uncontrolled climate change. The Department of Energy had its own assessment underway and was expected to claim it was too early to take action. Unable to come up with a consensus position, the agencies proposed kicking the problem upstairs to an intergovernmental level. The World Meteorological Organization and United Nations Environment Programme agreed. It was a promising proposal. Because climate change is a global problem, all nations had to be involved. Developing nations needed to be brought on board, but they were suspicious, if not openly hostile, regarding initiatives from the North. Being intergovernmental gave the IPCC credibility with government bureaucrats, making them more willing to come to the negotiating table.

At the same time though, the merging of science and politics undermined the credibility of IPCC assessments by making them vulnerable to the charge of political manipulation, an accusation successfully hurled at them by climate change deniers. Tying the scientific horse to the political cart was necessary, but also undermined the possibility of urgent action. Enough was already known about the science of climate change — ask the big oil companies — to proceed with the obvious solution: cut sharply and deeply into carbon dioxide emissions.

But this was as much a political as a scientific problem. Big Oil couldn't influence scientists except by countering their claims with its own claims, just as tobacco companies did. But oil companies could influence politicians. They were very good at it. By pushing scientific knowledge into the public realm, Big Oil ensured it would have more influence over the outcome, as it did when the American Petroleum Institute transferred its environmental unit into its political department. And by requiring the IPCC to go back to square one in its climate research, the American government and its corporate allies bought time. By the same token, because governments from around the world were being assembled into a single organization, the creation of a climate convention was made more feasible. The Framework Convention on Climate Change followed the formation of the IPCC by four years. The next challenge would be to pacify the framework convention.

CHAPTER 3
RIO TO KYOTO:
THE ROAD NOT TAKEN

The United Nations Conference on Environment and Development —
a.k.a. the Rio Summit or Earth Summit — was held in Rio de Janeiro in
June 1992. With 30,000 attendees, including more than 100 heads of state,
it was one of the largest such gatherings held to that point. It was also one
of the most important, since it established the framework for decades of
negotiations on climate change as well as on other significant treaties and
agreements. The year 1992 was chosen to mark the twentieth anniversary of
the UN Conference on the Human Environment that Maurice Strong had
overseen in Stockholm.

The world at Rio was very different from the world at Stockholm.
The Soviet Union was gone. The plan for globalization discussed by the
Trilateralists in the 1970s was a reality — the world's 500 largest corporations controlled some 70 per cent of world trade. But the world was the
same in other ways. Control over energy supplies still preoccupied the big
industrialized powers: the US under George H.W. Bush had recently invad-
ed Kuwait to protect Middle East oil. And Maurice Strong was still presiding
at the nexus of environment and business.

Strong had been a commissioner on the World Commission on
Environment and Development — the Brundtland Commission in 1987.

50

Brundtland's cure for the world's ills was more rapid economic growth in both industrialized and developing countries.[1] The commission called on the UN General Assembly to convene an international conference to transform its proposals into governmental undertakings and UN declarations. Strong and the other commissioners fanned out around the globe to spread the message. Strong was spotted in Davos in early 1989 co-chairing the World Economic Forum, where he put the greenhouse effect on the agenda — it cannot be resolved without the involvement of business leaders, he said — and brought in a phalanx of scientists to educate the business elite.[2] Then he showed up in his home country at the annual general meeting of the Federation of Canadian Municipalities in Vancouver, where he said that economic growth was essential but must be coupled with protection of the environment.[3] Then off to Ottawa later in the year to receive the Pearson Peace Medal, awarded to a Canadian who has contributed to peaceful change through law and Third World aid. Canadians should be ready for wartime-type sacrifices to solve environmental problems, he said, tailoring his message to suit the occasion.

STRONG AND THE BUSINESS COUNCIL FOR SUSTAINABLE DEVELOPMENT

Six months later, Strong was appointed secretary-general of the conference proposed by Brundtland. It would be held in Brazil as a gesture to the developing world. But the developed world and its corporate superstructure would be there too. Strong set up an office to oversee the conference, funded by the Rockefeller and Ford foundations, and some large oil and chemical companies based in the UK, Japan, Switzerland, Italy and the US.[4] Strong appeared at succeeding Davos gatherings promoting the conference. He turned up at the 1991 event with Stephan Schmidheiny in tow.

This Swiss billionaire industrialist (2018 net worth US$2.5 billion) inherited a vast sprawling industrial empire that produced house siding and roofing components using asbestos. Schmidheiny set out to remove that deadly material from his factories and replace it with a safer alternative. It was sustainability in action and Schmidheiny received plaudits for it. He gave a talk at one of the Rio preparatory meetings, where he was recruited by Strong to be his principal adviser on business and industry. Strong's rationale for privileging business over other sectors of society was that "industry is at the leading edge of these changes" to improve the living standards of people in the developing world. "There is a growing awareness that this is a key economic and business issue and that businessmen have to take the lead."[5]

And Schmidheiny did. He founded and initially financed the Business Council for Sustainable Development (BCSD). He followed the model of the Business Council on National Issues and the Business Roundtable, where only CEOs or top company decision makers could be members. This approach was meant to ensure rapid corporate buy-in without having to go back to corporate headquarters to get approval from higher-ups. Hence BCSD comprised forty-eight chief executives from some of the largest companies and worst polluters on the planet, such as Chevron, Volkswagen, Ciba-Geigy, Mitsubishi, Dow Chemical, DuPont and Royal Dutch Shell.[6] They had all benefited greatly from globalization and would let nothing stand in the way of free trade and open borders.

To cement his relationship with Strong, Schmidheiny recruited Canadian Hugh Faulkner as executive director. Faulkner had been a minister in the Pearson and Trudeau cabinets and had known Strong when he headed the Canadian International Development Agency and Petro-Canada. Faulkner left politics after a 1979 election defeat and spent the next eight years as an executive with Alcan Aluminium. He then was appointed secretary-general of the Paris-based International Chamber of Commerce, a global business lobby that came to play a leading role in climate change negotiations. Faulkner moved over to Schmidheiny's group two years later.[7]

Because of this relationship, BCSD enjoyed unparalleled access to the Rio conference secretariat and extraordinary influence in weakening the key agreements that came out of the conference. Schmidheiny and the council released their report, *Changing Course*, a month before the conference. It was one of the most influential reports to come out of the conference, which promoted a brand of market-led environmentalism and became a bestseller. Strong reported he co-edited a chapter in the book with DuPont's CEO.[8] Making profits and safeguarding the environment are not mutually exclusive, the book claimed. It argued for self-regulation rather than environmental protection laws. And it didn't set targets for reducing greenhouse gas emissions, an omission that provoked criticism from environmental groups. Greenpeace called it "a veiled attempt to minimize international environmental controls on big business . . . The shameful bottom line is that the BCSD, including as it does a number of oil companies, advocates the use of oil and gas reserves until they are exhausted," Greenpeace wrote in a statement. "How can business honestly say this is sustainable?"[9]

To ensure the corporate viewpoint would have maximum impact, Schmidheiny hired Burson-Marsteller, then the world's largest PR firm, with an unparalleled record of success in corporate greenwash. Nuclear power plant maker Babcock and Wilcox used the firm to restore its reputation after the Three Mile Island disaster in 1979. Burson-Marsteller assisted Union Carbide after the Bhopal disaster in 1984 and Exxon after the *Exxon Valdez* oil spill of 1989.[10] It worked for the B.C. Forest Alliance to attack environmental campaigns to save old-growth forest. The year after the Rio Summit, Burson-Marsteller masterminded a front group called the American Energy Alliance to defeat US president Bill Clinton's proposed BTU (British thermal unit) tax on fossil fuels. This tax was the centrepiece of Clinton's ambitious plan to combat global warming, which would reduce US emissions of greenhouse gases to their 1990 levels by the year 2000.[11] Its defeat undermined efforts to tackle climate change in the US for several decades.

Burson-Marsteller's missions for Schmidheiny were to greenwash the image of big business and to ensure that no binding regulations to control the activities of global business corporations were adopted. The combined BCSD-Burson-Marsteller efforts were so successful that proposals drawn up by the UN's own Centre on Transnational Corporations concerning the environmental impact of large companies — BCSD's own members, in other words — and issues of corporate responsibility and accountability were deleted from the Rio agenda following intensive lobbying by BCSD.[12] The best hope for imposing accountability on the activities of big business was lost. The centre had already been axed by the UN's new secretary-general, Boutros Boutros-Ghali. It was the beginning of a silent corporate takeover of the United Nations Framework Convention on Climate Change (UNFCCC) negotiation process.

Rio was "the largest and most costly diplomatic gathering in the history of the entire world," according to Sir Geoffrey Palmer, the former prime minister of New Zealand and an official Rio delegate.[13] So it had to have payoffs that could justify the enormous expense. At the end of the summit, more than 130 nations signed conventions on climate change and biodiversity. The delegates also reached agreement on Agenda 21, an action plan to develop the planet sustainably during the twenty-first century, and a broad statement of principles for protecting the world's forests. They also accepted the Rio Declaration, a non-binding statement of principles for environmental policy.[14]

The first report of the UN Intergovernmental Panel on Climate Change (IPCC) in 1990 stated that the only hope for avoiding unprecedented and ecologically disastrous global warming is to make deep cuts to carbon dioxide emissions.[15] Given this unambiguous statement, the UNFCCC that delegates signed declared that its objective was to "stabilize greenhouse gas concentrations in the atmosphere at a level that would prevent dangerous anthropogenic interference with the climate system."[16]

Yet it set no binding limits on greenhouse gas emissions for individual countries and contained no enforcement mechanisms. Participating nations pledged to voluntarily reduce their carbon dioxide emissions to 1990 levels by 2000. By the end of the year, 155 states had signed the convention. Who wouldn't? The corporate sector was confirmed as the key actor in "the battle to save the planet," *The Ecologist,* a radical British environmental magazine, concluded. "The best that can be said for the Earth Summit is that it made visible the vested interests standing in the way" of local communities seeking to reinstate their "social and political authority."[17] Maurice Strong had ensured that the global environmental movement would be dominated by business.

One major corporate success was the insertion of "joint implementation" into the 1992 UNFCCC. This allowed a country to meet its voluntary emission targets in part by offsetting its domestic emissions with projects — "offsets" — that reduced emissions in other countries. This possibility was of vital importance to companies such as Alberta-based TransAlta Utilities Corporation, which in 1993 generated 94 per cent of its electric energy from coal. As one industry observer commented, "If government regulations were to come down overnight, TransAlta would be hit hard. That's why they're seriously looking at offsets."[18] TransAlta headed the BCSD's joint implementation project. Under joint implementation, it could invest in lower-emitting electric utility systems in India, for example, and receive emission credits even though its coal-fired power plants continued to emit as much greenhouse gases as ever.

THE WORLD BUSINESS COUNCIL FOR SUSTAINABLE DEVELOPMENT TAKES OVER

Impressed with BCSD's successes at Rio, member CEOs wanted to keep the momentum going, continuing to oppose enforceable environmental legislation and pushing for voluntary corporate action. In 1995 BCSD merged with the World Industry Council for the Environment to create the World Business Council for Sustainable Development (WBCSD). The former was another organization of CEOs created by the International Chamber of Commerce to promote the industry agenda after Rio.[19] It would be a key participant in conferences of the parties that resulted from the UNFCCC.

The International Chamber of Commerce, with some of the same corporate members as the BCSD, had developed its own business charter on sustainable development. Chair of the committee that produced the charter was Peter Bright, head of environmental affairs at Shell International. (BP, Texaco and Mobil also signed it.)[20] This document committed its members to sustainable *economic* development. What this meant was, in fact, escalating economic development, because "economic growth provides the conditions in which protection of the environment can best be achieved," as the chamber's executive board put it.[21] Underlying the charter's lofty rhetoric was a requirement that all action be voluntary and rely not on environmental laws and regulations, but on pressure from customers, employees and other companies to make firms toe the line, an approach that lacked credibility for most environmentalists.[22]

A Greenpeace International study identified the key role WBCSD played in preventing effective climate legislation. Its strategy was to move from stakeholder observer role, one played by many sectors of society such as farmers, Indigenous peoples, research organizations, trade unions, local governments and environmental NGOs, to a more active and institutionalized role, thus increasing its influence. With so much on the line, fossil fuel companies took an active role in WBCSD's executive committee. Its first chair was the managing director of BP; Royal Dutch Shell's chair took over the helm of the organization several years later.[23] And the influence continued. In 2015 Marvin Odum, Shell's US country chair and upstream Americas director, was WBCSD's vice-chair.

GLOBAL CLIMATE COALITION PROMOTES DENIAL

While Maurice Strong and the WBCSD laid the ground for reframing global warming as an issue requiring a business solution, business implemented a strategy of denial to buy time to remove regulation from the negotiating table and to transition to new areas of economic activity. "The purpose of science denialism has been to quash all immediate impulses to respond to the perceived biosphere crisis," by seeding the marketplace of ideas with doubt and confusion, philosopher of economic thought Philip Mirowski explains. "Denying the very existence of global warming is cheap and easy to propagate, can be fostered quickly, and tends to draw attention away from issues of appropriate responses to crisis," he writes.[24]

Industry's first foray into denialism was through the Global Climate Coalition, which was established in response to the formation of the United Nations Intergovernmental Panel on Climate Change (IPCC) in 1988. About fifty companies in the mining, oil and gas, electricity, automobile and chemical industries joined the effort, but ExxonMobil and the American Petroleum Institute paid most of the bills. Global Climate Coalition's mission was to convince the US Congress and the media that the idea of human-caused global warming was not true.[25] The coalition kept the discussion focused on whether or not there was a problem, thus preventing discussion about what should be done.

The Global Climate Coalition brought back Bruce Harrison, the prominent public relations strategist on environmental issues who had led the attack on Rachel Carson in the 1960s. Harrison spent the intervening decades practising anti-environmental public relations. He led the RJ Reynolds-sponsored Total Indoor Environmental Quality coalition, whose mission was to blur the impact of second-hand smoke in enclosed spaces.[26] He used many of the tactics from the Rachel Carson campaign to discredit both the IPCC, whose reports were documenting the reality of global warming, and environmental groups, who argued that greenhouse-gas emissions had to be controlled. The Global Climate Coalition strategy was to spread the notion that global warming was a dangerous myth and to emphasize the lack of absolute proof.[27] It promoted its denial message as the agenda for Rio was being finalized. At the last session of pre-conference negotiations, the Global Climate Coalition distributed flyers claiming, IPCC evidence notwithstanding, that there would be no environmental benefit from stabilizing carbon dioxide emissions.[28]

Yet while the Global Climate Coalition denied the reality of climate change in all public venues, privately its members were well aware it was happening. As we've seen, as early as the 1970s most Big Oil companies had reports from their own scientists detailing increased concentrations of greenhouse gases in the atmosphere. Scientists were even preparing reports for Global Climate Coalition members. A leaked internal memo, "Predicting Future Climate Change: A Primer," was written in 1995 by a Mobil Corporation scientist for distribution to coalition members. It stated "the scientific basis for the Greenhouse Effect and the potential impact of human emissions of greenhouse gases such as carbon dioxide on climate is well established and cannot be denied."[29] They denied it anyway. Exxon, soon to be united with Mobil to become the largest oil major, continued funding denial efforts at least until 2008, donating over US$30 million to think tanks and astroturf groups. These are "apparently grassroots-based citizen groups that are primarily conceived, created and/or funded by corporations, industry trade associations, political interests or public relations firms," according to the progressive watchdog group, the Center for Media and Democracy.[30]

As the Global Climate Coalition forged ahead with its message, a second source of denial was bulking up. Big Oil was impressed by the successful campaign waged by a seemingly independent network of neoliberal think tanks to manufacture doubt about the link between second-hand smoke and lung cancer. By the 1990s, neoliberalism had extended its influence to every corner of the globe and was most firmly entrenched in the UK and the US. This campaign to discredit the science around smoking and cancer was successful because it delayed government action for several decades, allowing tobacco companies to continue reaping their gargantuan profits and enabling them to transition to lucrative markets in Russia, Asia and Latin America.

Even though global warming and second-hand smoke occurred on vastly different scales, for neoliberals they shared a common problematic: There was a prevailing belief in society that these issues needed to be solved through government intervention. The state had a crucial role to play for neoliberals, but it was to create and enforce markets, not intervene in them. If neoliberalism could have been reduced to a slogan, it would be "all problems must be solved by the market." In fact, there were no problems in society; it's just that the market hadn't been given the opportunity to operate properly.

With the financial support of fossil fuel companies and the foundations of a dozen conservative billionaires, many of these think tanks turned to global warming denial in the 1990s. They called themselves skeptics and contrarians, but neither of these terms applied. "'Skepticism' suggests an open mind," Australian political scientist Robert Manne explains. But "the minds of those who dispute the consensual core of climate science are closed." And as for contrarian, it "might be the right term for the small minority among climate scientists who have not accepted the consensual conclusion of their fellow scientists," Manne writes. "Most of those who dispute the consensual conclusions" of climate scientists are not mavericks or heretics but "orthodox members of a tightly knit group" who disseminate ideas, first developed by the Mont Pèlerin Society. They are cogs in a highly organized denialist machine.[31]

The deep pockets of their funders fuelled a twenty-five-year campaign to manufacture doubt and undermine scientific evidence about the link between carbon dioxide emissions and global warming. By the early 2000s, climate scientists were virtually united in their view that global warming was real, was caused by human activity and will have — and was already having — disastrous consequences for the planet. The denial campaign was astonishingly effective in confusing the public and politicians about the reality of global warming and delaying meaningful responses from most governments.[32] This was not a problem for neoliberals, for whom lying and spreading falsehoods about global warming was as valid as messages disseminated by scientific experts, Philip Mirowski explains, "[B]ecause the final arbiter of truth is the market, and not some clutch of experts who represent sanctioned science."[33]

The neoliberal assault on action designed to mitigate global warming emerged in 1990, in tracking the work of the IPCC.[34] Only a handful of denial publications was produced over the next five years, as industry was putting its denialism money into the Global Climate Coalition. Publications took off in 1996 and 1997, the year of the Third Conference of Parties in Kyoto, Japan, where the threat of internationally binding action loomed. (The Conference of Parties is the decision-making body responsible for monitoring and reviewing the implementation of the United Nations Framework Convention on Climate Change.) More publications were produced in 1996 than in 1990–95 combined, and the output in 1997 was more than five times that of 1996. It was a highly co-ordinated effort to halt the United States endorsement of the Kyoto Protocol, even though the US government had faithfully followed market imperatives by successfully inserting carbon markets into the agreement.

Over the seven-year period (1990–1997) six think tanks produced 77 per cent of the books, op-eds, articles, policy studies and news releases attacking global warming.[35] The news release was pressed into service in the crucial year of 1997 to have more direct impact on public opinion. Five of these think tanks were created within three years of one another in the early 1980s, during the scale-up in the neoliberal infrastructure. They were the National Center for Policy Analysis (Dallas, Texas), Heartland Institute (Chicago, Illinois), National Center for Public Policy Research (Washington, D.C.), Competitive Enterprise Institute (Washington, D.C.) and the George C. Marshall Institute (Arlington, Virginia). Their original mission, obviously, was not to attack global warming, since they predate its rise as a public issue. But they quickly turned their guns in that direction as fossil fuel and foundation funding poured into their coffers. The sixth think tank, the Hoover Institution on War, Revolution and Peace, located at Stanford University, is the granddaddy of such organizations. It was created in 1919 by US president Herbert Hoover to counter the spread of the 1917 Russian Revolution, three decades before Hayek and Friedman established the Mont Pèlerin Society. The Hoover was therefore not originally neoliberal, but it soon entered the fold. By 2010 Hoover housed ten Mont Pèlerin Society members, one of the highest concentration of any organization in the world.[36]

Several observations have been made about these publications over the initial seven-year period. First was the correlation between the release of Intergovernmental Panel on Climate Change (IPCC) reports and the occurrence of other significant events and the scale-up in denialist publications, demonstrating a rapid-response capability. Second, these organizations had access to large pools of corporate and foundation funding. The sources most commented on have been ExxonMobil Corporation and the various foundations of Charles and David Koch, multi-billionaire owners of Koch Industries, the second-largest privately-owned company in the US with extensive oil and gas operations. A third observation is that the think tanks provided funding and organizational support for a handful of scientists "skeptical" about the reality of global warming. These scientists, many of whom have no expertise in climatology or related disciplines, wrote most of the studies, reports and books the think tanks published.

Three messages stand out in the publications. The most frequent by far is the claim that the scientific evidence for global warming is weak and even wrong. It is "contradictory," "flawed" and "murky." There is no consensus, the publications falsely allege. Some publications charge that global warming is an attempt by radical environmentalists to take over the world. A second key claim is that proposed action will do more harm than good to the economy, using words such as "devastating," "staggering" and "crippling." A third claim is that global warming, if it is occurring, will be beneficial because a moderately warmer climate will be a far better one for humanity. In total, these claims lead to the conclusion that no action is warranted. Uncertainty over climate science was tied to certainty over economic devastation to create a powerful propaganda message that succeeded in obstructing national climate policies, particularly in the US and the UK, and international climate change mitigation agreements.

SECOND, MARKETS

For the first few years, the Global Climate Coalition, the World Business Council for Sustainable Development, the International Chamber of Commerce and allies like the American Petroleum Institute succeeded in averting mandatory emission reductions. But at the first Conference of

Parties in Berlin in 1995, the environmental movement, acting mainly through the Climate Action Network, was able to persuade the conference to ask governments to adopt quantified emission reduction commitments by 1997.[37] Buttressing this position was the IPCC's second assessment report, created by the UN's panel of 2500 scientists from around the world. This report concluded that global warming was occurring and presented dangers to society. As *The New York Times* reported on its front page, "The earth has entered a period of climatic change that is likely to cause widespread economic, social and environmental dislocation over the next century if emissions of heat-trapping gases are not reduced, according to experts advising the world's governments."[38]

Doing something about this situation gained in urgency. Some companies, led by BP and DuPont, broke from the Global Climate Coalition. In a highly publicized speech that also received front-page treatment in *The New York Times*, BP chair John Browne said that the time to contemplate action "is not when the link between greenhouse gases and climate change is conclusively proven, but when the possibility cannot be discounted and is taken seriously." British Petroleum "has reached that point."[39]

BP and DuPont joined forces to advocate for markets as the solution to global warming. Like other business interests, though, BP continued to oppose mandatory emission limits and to resist carbon taxes, at least until business pivoted to carbon pricing and clean growth about a decade later. If we must have mandatory emission limits, the reasoning seemed to go, then we would meet them through market mechanisms and not government regulation. BP and DuPont joined the International Climate Change Partnership (ICCP), a business group composed largely of chemical companies that participated in the negotiation of the Montreal Protocol in 1987, an international treaty designed to protect the ozone layer.

Jonas Meckling, a climate policy analyst at the University of California at Berkeley, argued that the International Climate Change Partnership "was the first business organization to advocate market mechanisms as part of a climate treaty."[40] The Global Climate Coalition would lose much of its influence when both BP and Shell exited the group. It disbanded in 2002 after President George W. Bush and Congress rejected the Kyoto Protocol, with its mandatory cuts, and withdrew American support. The organization explained "it had served its purpose by contributing to a new national approach to global warming."[41]

Just as American-based ExxonMobil led the Global Climate Coalition and other denial forces, so the British-based BP came to lead the International Climate Change Partnership and pro-market forces. The two companies covered the range of allowable thought: global warming isn't occurring, so we must do nothing to upset the current economic system; but if it is occurring we must deal with it through market mechanisms. Both strategies were designed to undermine society's ability to deal directly with environmental problems through government regulation. Both strategies derived from neoliberal roots, denialism originating in the work of Mont Pèlerin Society founder Friedrich Hayek, while market solutions to environmental problems can be traced to Mont Pèlerin Society lifetime member and University of Chicago economist Ronald Coase. "Political actors originally bent upon using state power to curb emissions directly are instead diverted into the endless technicalities of instituting and maintaining novel markets for carbon permits and offsets while carbon emissions grow apace," wrote Philip Mirowski and his colleagues in the Australian journal *Overland*.[42]

The question is often posed as to why the European Union shifted from being in favour of direct regulation and staunchly opposed to emissions trading, to becoming the major arena for emissions trading. The answer, according to Meckling, is BP. That company played a key, if not leading, role in bringing markets to the UN Framework Convention on Climate Change and Kyoto. BP representatives chaired both the World Business Council for Sustainable Development and the ICCP. BP partnered with the American NGO Environmental Defense Fund (EDF) to design an internal emission trading scheme for its own use, anticipating that it would create a positive frame for "emissions trading and thus forestall alternative responses such as an emissions tax, which was perceived to be more costly," Meckling observes.[43] The EDF created a narrative that carbon markets work and are the best solution to climate change.

In 1988 the EDF persuaded US President George H.W. Bush that a market for sulphur dioxide emissions was the cheapest way to deal with acid rain. In 1995, it persuaded Bill Clinton, Bush's successor and supposed political opponent that market mechanisms were the way to deal with global warming. The EDF promoted and helped design markets for sulphur dioxide to reduce acid rain. This worked, conventional wisdom has it, and became the success story that Clinton brought to the second Conference of Parties in Geneva in 1996.

BP had already persuaded the British government to accept the emissions-trading scheme, not a difficult task because the government already leaned in that direction. The UK held the EU presidency and negotiated on behalf of the EU, so a compromise was not long in coming.[44] The US threatened to pull out if emission trading wasn't included in the negotiations. It was, and the sting of mandatory emission limits was removed.

Before the third Conference of Parties in Kyoto, Bill Clinton said he aimed to "harness the power of the free market" to mitigate greenhouse gases.[45] "If we do it right," he continued, "protecting the climate will yield not costs but profits, not burdens but benefits, not sacrifice but a higher standard of living." Clinton was almost evangelical in his belief in the miracles carbon markets could deliver to humanity. So was Vice President Al Gore, who led the American delegation to Kyoto and brokered the compromise with the EU. The International Climate Change Partnership advocated for this deal, still opposing mandatory reduction targets contained in the Berlin mandate, which it called "impractical short-term targets for greenhouse gas emissions reduction" and supporting US proposals for "flexible, cost-effective and fair approaches to greenhouse gas emissions reductions that will ensure economic prosperity."[46] The deal was done.

CHAPTER 4
TOM GETS HIS WAY; MURRAY ENTERS THE FRAY

That the Business Council on National Issues (BCNI) became such an influential voice in Ottawa was in no small part due to the drive and persuasive power of one man — not one of its CEO members, but its president, Tom d'Aquino. And he was of the same view as Maurice Strong that business had to take the lead in dealing with environmental issues. He set out the work plan for his organization's management of the environmental problem at a meeting of business and environmental leaders at the Americas Society headquarters on New York's Park Avenue in December 1989. If corporate leaders did not seize the initiative and define environmentalism, he said, "others will, and business will have to live with the results."[1]

The CEOs went right to work. Their first foray into global warming politics was made behind the scenes in 1990 when the Mulroney government's Green Plan was being formulated. A draft section calling for a carbon tax was leaked to the CEOs and they immediately called the Prime Minister's Office to object and demand to be more actively consulted. What was said is not known, but the final draft of the plan did not include this provision, to the relief of the fossil fuel industry.[2]

The organization took a more pre-emptive role in the preparations for the Rio Summit. BCNI vice-president John Dillon was lead business representative on the Canadian delegation to the summit, giving him unique access to federal Environment Minister Jean Charest and senior government officials.

In preparation for the summit, BCNI established a task force on sustainable development. Three of its six members came from that font of sustainability, the fossil fuel sector (Shell Canada, Imperial Oil, TransAlta Utilities).[3] After the summit, as we've seen, the Business Council for Sustainable Development (BCSD) was transformed into the World Business Council for Sustainable Development (WBCSD), to bring the business viewpoint to succeeding conferences of the parties. The close connection between BCNI and the global body ensured that business could co-ordinate its efforts at national and international levels, a phenomenon that would be observed in 2007 when both organizations pivoted from resistance to almost any action on climate change to advocacy for clean growth and carbon pricing. Interlocking directorships helped facilitate interaction between the organizations. Noranda CEO David Kerr, for instance, was a member of BCNI's policy committee and vice-chair of the WBCSD.[4]

Canada came out of Rio with a commitment to stabilize greenhouse gas emissions at 1990 levels by 2000. D'Aquino worked with the Canadian Association of Petroleum Producers (CAPP), the industry's most powerful lobby group, to persuade federal and provincial governments to institute a voluntary approach to reductions. CAPP included all up stream operators plus financial services firms, law firms and drilling companies as associates. Its first president, Gerry Protti, was a former assistant deputy minister in Alberta Energy and also a vice-president of EnCana Corporation. CAPP's second president was David Manning, who transitioned into this job from his post as deputy minister of Alberta Energy — where he'd been Protti's boss — highlighting the links between government and industry in Alberta. It was Manning who came up with the voluntary emissions plan.[5]

"We believe that a voluntary industry program must form a cornerstone of Canada's national action plan on climate change," the BCNI's task force on the environment declared in a 1994 report. "We are convinced that regulation alone is not the answer to address an issue that has such widespread implications for business and consumers." Like other council bodies, this task force was dominated by the fossil fuel industry, with eight of fourteen members — including two of three co-chairs — drawn from that sector.[6]

CAPP was right behind the CEOs. "We will commit to have our members report on a regular basis their emissions and to demonstrate their ability to show improvements," Protti pledged.[7] The aim was to show government and the public that industry could cut its emissions without prescriptive regulations. Chrétien's Natural Resources minister, Anne McLellan — his senior, and sometimes only, MP from Alberta — enthusiastically supported the proposal, asking companies to develop their own plans to limit or reduce greenhouse gas emissions and record the results in a voluntary challenge and registry program.

The program was called the National Action Program on Climate Change and was unveiled at the first meeting of the UN Framework Convention on Climate Change (UNFCCC) in Berlin in 1995.[8] "Good corporate citizenship and social responsibility is driving us forward," d'Aquino effused at a news conference organized jointly by Natural Resources Canada and the BCNI.[9]

But good corporate citizenship and social responsibility soon stalled, according to studies of the registry by the Pembina Institute, a leading clean-growth think tank. In one study released in 2002, after the registry had been operating for seven years, Pembina found a fatally flawed program in which many companies had not registered at all, most companies that did register failed to report their emissions, and most of those that did report saw their emissions "increase significantly." Summarizing his research, Pembina's Matthew Bramley wrote: "In 1992, Canada signed an agreement to stabilize emissions at the 1990 level by 2000. Instead of mandating reductions, the government requested voluntary compliance by industry. The result has been a 24 per cent increase in total industrial emissions above the 1990 level."[10] Bramley's conclusion was unequivocal: "unilateral voluntary approaches have failed. They are thoroughly discredited by the results presented in this report. The only credible way for Canada to address climate change is through the [mandatory] Kyoto Protocol."[11] (This was the binding international treaty signed at Conference of the Parties 3 in Kyoto, Japan, in 1997. Developed nations that were parties to the UNFCCC committed to reduce their greenhouse gas emissions 5 per cent below 1990 levels between 2008 and 2012.)

This was not industry's conclusion. As Kyoto loomed with its proposed binding limits, BCNI and CAPP went all out to forestall its progress, arguing that such limits could devastate Canada's growth prospects. The agreement to curb global warming could pose a worse threat to the Canadian economy than Quebec secession or the country's $600 billion debt, BCNI warned.[12] Canada should not have to meet the same requirements as other countries because it relies so heavily on oil, coal and gas, the council argued. BCNI also wanted all countries to be governed by the caps, even though developed countries like Canada were overwhelmingly responsible for the high levels of carbon dioxide in the atmosphere.[13] Six months before the Kyoto meetings, d'Aquino drafted a letter, signed by 100 CEOs — including executives from Imperial Oil and Shell Canada — warning that even that modest greenhouse gas reduction goal could knock 3 per cent off the country's GDP, killing jobs and harming exports.[14] Elizabeth May, then president of the Sierra Club of Canada, called the BCNI claims an exaggeration, designed perhaps to ensure no new binding agreement was reached at Kyoto.[15]

Nonetheless, business had good reason for believing it was in the driver's seat. In the fall of 1997, for instance, when d'Aquino gave a presentation on what Canada should strive for at the Kyoto Conference, deputy ministers from seventeen federal departments gathered to watch his slide presentation — a record rarely, if ever, surpassed. D'Aquino could outmanoeuvre any bureaucrats who dared take him on, business journalist Peter Newman observed, by the simple gambit of being smarter and faster than they ever dared to be.[16]

CHRÉTIEN COMMITS TO KYOTO

A split was occurring in the fossil fuel industry that sapped the industry's efforts to present a united front in its PR campaign to undermine mandatory caps on emissions. At the international level the split appeared, as we saw in the previous chapter, when BP CEO John Browne declared that global warming was real and action must be taken. BP withdrew from the Global Climate Coalition and Royal Dutch Shell followed the next year. A similar split occurred among the members of CAPP, where president Manning promoted co-operation with federal regulators. He was supported by three large oil patch players, Suncor Energy, Shell Canada and Petro-Canada (still partly owned by the Canadian government). But he was resisted by other major players that included Imperial Oil, Talisman Energy and Alberta Energy Corporation (later EnCana Corporation).[17] The split gave Chrétien room to manoeuvre. His government was torn between what the provinces and domestic industry wanted and what other nations, especially the US, were proposing. Chrétien wanted to be able to say he was doing more than the Americans to fight global warming without alienating Alberta.

He went down to the wire before announcing his government's target to reduce emissions by 3 per cent below 1990 levels by 2010. By this time Canada's emissions were already 13 per cent above 1990 levels. Tom d'Aquino was first off the mark, predicting that the federal government's target "will be difficult to achieve" and "will have wide-ranging implications outside Alberta too," as it hit Ontario manufacturers.[18] The Canadian delegation, however, was swept up in the negotiations and in the end accepted a goal of reducing greenhouse gas emissions by 6 per cent below 1990 levels by 2008–12.

At first, the fossil fuel industry didn't see this as an immediate threat to its business model because it was already operating under the ineffective five-year voluntary program established in 1995. (The program would be renewed for another five years in 2000 despite its failure to prevent steadily rising emissions, as the Pembina Institute noted.)[19] And it was confident it could dissuade Chrétien from ratifying Kyoto after he saw how damaging the accord would be to the Canadian economy. It could point to his hypocrisy in committing to emission reduction at Kyoto while at the same time enthusiastically backing Alberta oil sands development.

THE KLEIN YEARS

March 6, 1996 was "a beautiful day for bitumen," industry magazine *Oilweek* effused.[20] Not only was there no carbon tax in Finance Minister Paul Martin's budget, tabled that day in the House of Commons, but the industry had received what it wanted — a universal low-tax regime for all oil sands producers. Before, Ottawa had distinguished between conventional mining projects such as Syncrude, Suncor and Cold Lake, and in situ projects.

Only about 20 per cent of the bitumen resource could be accessed through mining. The rest had to be tapped via wells. The usual technique was to drill twin wells, push pressurized steam down one well to soften the bitumen, which then is forced up through the second well. The tax regime had favoured mines by allowing a wider range of deductions. Now all projects would get the same generous deductions. This was crucial because most new projects would be of the in situ variety.

The measure accelerated tax writeoffs for capital investment — a major concession, given the massive capital investment required to bring a new project into operation. It was welfare for capitalists. "I really applaud the decision of the federal government," said Eric Newell, president of Syncrude Canada. "This is the announcement of the decade and takes Fort McMurray into the next century," the mayor of that soon-to-boom town enthused.[21] The tax giveaways were especially noteworthy because the Liberals were in the throes of fighting government deficits and debt.

The oil patch liked Martin's tax breaks so much it invited him to speak to the 1,600 members at CAPP's annual dinner in Calgary. It had been fifteen years since a Liberal finance minister addressed leaders of Alberta's oil patch and rarely had a federal Liberal received such a welcoming response.[22]

The federal tax breaks were the icing on the cake. Several months earlier, Ralph Klein's government in Alberta had enacted a breathtakingly low generic oil sands royalty regime that saw oil companies pay a nominal royalty of 1 per cent until developers recovered all project costs and then would start paying 25 per cent of net revenues. The definition of "all project costs" was generous. "The bulk of royalties will be shifted to future years," Alberta Energy explained.[23] It was an enormous subsidy from the owners of the resource to the industry, something Peter Lougheed would have never countenanced.

But Lougheed was no longer in office. After he left, neoliberalism crept in the door with Don Getty and slammed it shut with Ralph Klein, who rode into office in a blitzkrieg attack on the social democratic state and government itself. It didn't take long for market thinking to drown out all other ideas about the oil economy. Alberta — and Ottawa — was finally on board with the Thatcher-Reagan revolution. Klein cut his government from twenty-six to seventeen ministries and slashed $700 million in spending, committing to a balanced budget within three years. His government withdrew loan guarantees worth billions from companies the government had supported before, while it began to divest itself of shares in various industries it had backed. Klein sold the province's remaining stakes in Alberta Energy Corporation and Syncrude.

To advance the Klein revolution, the Fraser Institute opened a branch office in Calgary and staffed it with members of the Calgary School, the phalanx of neoliberal and neoconservative professors from the political science, history and economics departments at the University of Calgary.[24] They published studies on reducing debt, cutting taxes, privatizing health care, cutting social assistance and privatizing the Canadian Wheat Board, providing ersatz academic backup for Klein's radical restructuring of Alberta society.[25]

The climate the institute helped create in Alberta must have been a factor in Klein's multiple election victories, considering the deep cuts he made to social programs. At the Fraser Institute's thirtieth anniversary celebration in Calgary in 2004, then *Calgary Herald* editorial writer and future Wildrose Party leader Danielle Smith introduced Klein as the institute's "teacher's pet" for adopting so many of its ideas.[26] The institute showed its appreciation for his slash-and-burn approach to governance by creating an award for him, the International Fiscal Performance Award, and appointing him a senior fellow of the institute after his retirement from active politics.

As for the oil sands, the Canada-US Free Trade Agreement, with its proportionality chapter was in place and Jean Chrétien had approved the North American Free Trade Agreement. The oil and gas industry prepared to export record volumes of crude oil and natural gas to US markets and abandon Central and Eastern Canada, leaving those markets to fend for themselves. The National Energy Program (NEP) was long gone, replaced by corporate energy strategies, as an integrated North American energy market was on the ascent.[27] A sea change was occurring in how government handled the oil sands industry, which was becoming the leading driver of energy exports. When Klein became premier, oil sands production was 375,000 barrels per day. When he stepped down in 2006, it was 1.1 million. Alberta would no longer participate directly in oil sands projects. Instead it would create an investment climate to encourage money to flow into the province and projects to commence, just as Friedrich Hayek of Mont Pèlerin and his colleagues would advise.

NATIONAL OIL SANDS TASK FORCE

Klein had been in office for over a year when the oil industry set up the National Oil Sands Task Force to answer the question: What can we do to get things moving here? It was billed as a "collective of oil industry and government representatives," but the task force was nothing of the sort. Oil industry dominance was evident everywhere. The idea for such a body came from Eric Newell, head of oil sands giant Syncrude and the Alberta Chamber of Resources, the leading industry lobby group at the time (soon to be replaced by CAPP). The task force was comprised of a chair from Occidental Petroleum and six subcommittees, composed of fifty members — five from the federal government, three from the Alberta government and forty-two from the oil sands and supporting industries. There were no representatives from environmental or community groups or First Nations.

Syncrude had its own reasons for initiating the effort. The company had signed a royalty deal with Peter Lougheed in 1978 that was set to expire in 2003. Lougheed had negotiated a "pretty good deal for the Alberta government, I think it's fair to say," comments Paul Precht, an executive director from Alberta Energy, who was involved in the meetings from day one. Lougheed was "getting a lot of royalty dollars out of Syncrude." The company was motivated to get better royalty treatment on the expiry of the agreement and wanted any measures developed by the task force to be grandfathered to include the lowered royalty.[28] Given the task force's make-up, anything it came up with would be better than the agreement Syncrude had signed with the Lougheed government. Suncor also had a stake in the negotiations. It was mainly an oil sands miner, but had an in situ site next door that it wanted to develop.

The National Oil Sands Task Force's 1995 report, "The Oil Sands: A New Energy Vision for Canada," laid out a twenty-five year strategy that predicted oil sands production doubling or tripling to between 800,000 barrels per day and 1.2 million barrels per day by 2020. (That target was almost reached by the time Klein left office in 2006.) It also declared "oil sands" to be the official name for the bituminous deposits, aiming to boost acceptance among a public that had also been hearing the dirtier-sounding term, "tar sands." The oil sands, it said, were a "national prize."[29]

The task force had no difficulty getting through to the Klein government, thanks to Klein's energy minister Patricia Black, who was a perfect fit for the job. Before going into politics (and after leaving it) she worked in the industry, including a stint at Suncor. And, unlike Lougheed, she gave industry the lead in developing the royalty rate. The task force benefitted from the presence of David Manning, CAPP president and former deputy minister in Alberta Energy.

Key to getting through to Chrétien and Martin was Natural Resources Minister Anne McLellan.

"Landslide Annie" won her Edmonton seat in the 1993 election by one vote — or rather, twelve after a judicial recount. Chrétien immediately appointed her to head the energy ministry as a message of reassurance that things had changed since the bad old days of the NEP, even though he had been Trudeau's energy minister during most of the NEP period (1982–1984). McLellan worked hard and successfully to gain industry's trust on the tax and Petro-Canada privatization files. She became a serious oil sands booster. "There is no doubt in my mind . . . the oil sands can be the Leduc of the next fifty years," she told the Edmonton Chamber of Commerce as the Liberals were working on their oil sands tax incentive proposals.[30]

To augment its ability to reach Chrétien and Martin, CAPP hired Chris Pierce, who had run and narrowly lost as a Liberal in 1993 in an Edmonton riding adjacent to McLellan's. He was appointed CAPP's vice-president of government relations. He came through, arranging a meeting between Martin and CAPP's president, David Manning, that got the ball rolling.

"The agenda for the second generation of oil sands producers was completed by 1997," writes Alberta historian Paul Chastko.[31] That same year, the Conference of Parties agreed to the Kyoto Protocol that set mandatory limits on greenhouse gas emissions. It wouldn't take long for oil sands expansion and emission limits to meet head on.

It was not as if the industry wasn't fully aware of the consequences of Kyoto: As president of CAPP, Manning was a delegate to Kyoto. Albertans received a taste of the coming conflict when Jean Chrétien flew into Fort McMurray by jet and military helicopter to announce $5 billion worth of new oil sands projects. "It's fantastic because we have more oil here than Saudi Arabia," he told 1,800 cheering residents, oil industry executives and politicians packed into the local curling rink. He next went to the Syncrude site for a photo-op, sitting in the driver's seat of one of the world's largest graders as hundreds of employees applauded.

But the Pembina Institute was on Chrétien's trip, too. It warned that the new investment "threatens to make a mockery of Canada's commitments to fight climate change." If the government was serious about protecting the environment, it had to stop these expansion plans.[32] But that, apparently, was for the future. The day Canada signed the Kyoto Protocol, *The Globe and Mail* reported that $20.6 billion worth of oil sands projects were on the books for the next decade. That included a $6.6 billion expansion for Syncrude and $2.8 billion for Suncor Energy, two immediate beneficiaries of the new pricing regime they helped bring to fruition.[33]

MURRAY EDWARDS MAKES HIS NAME

Yet the stampede of new oil sands production was led not by the two original big oil sands firms but by Calgary oilman Murray Edwards, who seemed to have a special knack for exploiting the new tax and royalty regime.

In the Trudeau-to-Trudeau time frame, Edwards emerged as a private sector opposite number to Maurice Strong. Where Canada's energy riddles propelled Strong to the peak of global politics, they made Murray Edwards very rich in a strictly Canadian setting. The latter man's story is inseparable from the rise of the oil sands in both Canadian business and Canadian politics.

The country's ultimate business chronicler, Peter Newman, said that Edwards revolutionized the way wealth was created in the oil patch. Edwards was at the hub of Calgary's most significant business network and his every move was discussed at the Petroleum Club and other venues. His genius, Newman says, was his timing. Newman compared Edwards to Wayne Gretzky. Like Gretzky did on the ice, Edwards could see plays developing and would be precisely at the right spot at the right moment. It's a talent that turned a $100,000 stake into something worth nearly $1 billion in just over ten years.[34]

Edwards arrived in Calgary in 1983, a twenty-four-year-old graduate from the University of Toronto law school. He had already earned a commerce degree from the University of Saskatchewan. Law plus commerce: the foundation for a promising business career.

THE BIG STALL

He went to work at Burnet Duckworth and Palmer, Calgary's top law firm, with a stable of oil and gas company clients. Within four years, Edwards became a partner. He saw oil prices collapse in 1986 and witnessed the carnage in corporate boardrooms. One of his major accounts was Peters and Co., an influential fossil fuel investment firm. He crossed from law to business when he and some law firm colleagues set up a firm, together with Peters, to invest in oil and gas companies that were financially weak yet presented investment opportunities because of their assets or tax pools. The time was right for a shrewd investor to enter the fray.

Because of low oil prices and the scant movement on the oil sands front, Big Oil was stepping back from Western Canada. Edwards was one of the first to see future potential. "There was a vacuum," he told *National Post* energy columnist Claudia Cattaneo. "I was fortunate to fit into that vacuum."35

Still in his twenties, Edwards invested $100,000 to help recapitalize a near-dead company called Canadian Natural Resources. The company was heavily in debt due to excessive leveraging in the late 1970s and 1980s, when oil prices were at record levels. After the 1986 oil price collapse, the company was forced to sell off most of its assets. A distressed Canadian Natural Resources was the kind of company Edwards went after.

Canadian Natural Resources was worth about $1 million when Edwards and partners took over. The company had four employees and produced one hundred barrels of oil a day. The stock traded at ten cents a share.36

Edwards started out small, continuing to sell assets to pay off debt and using some of the proceeds to buy more promising properties. Within a year, shares jumped to $1.90 and production was up to 450 barrels a day. A year later, shares were up another 62 per cent. By 1992, production was 3,400 barrels a day and gas production 100 million cubic feet a day. The company continued to expand rapidly in conventional oil and gas production, and was able to buy troubled oil and gas producer Sceptre Resources for $715 million in 1996.

Edwards and his partners recapitalized two other distressed companies that would grow into multi-billion-dollar undertakings, Penn West Petroleum and Ensign Energy Services.

Canadian Natural Resources picked up some heavy oil properties near

Fort McMurray in the early nineties. They were cheap because the future of the oil sands was in doubt at the time.

But then, Ralph Klein and Jean Chrétien presented their gifts to the oil sands industry. Edwards was a well-known Liberal but was a friend of both men: He co-chaired lunches for Klein's Progressive Conservative party and two dinners for Chrétien.

The personal, political and business stars were aligning. After the tax and royalty changes, Canadian Natural Resources began amassing sections of land in the Pelican Lake area near Fort McMurray, buying leases from Koch Exploration Canada, a Canadian subsidiary of Koch Industries, the massive oil and gas empire owned by the Koch brothers. By 1997, Canadian Natural Resources had $800 million worth of oil sands projects in the works.

As his wealth grew, Edwards started diversifying, always in Canada. In 1994 he became part owner of both the Calgary Flames and miner Imperial Metals; and in 1995 he took a stake in Magellan Aerospace.

Edwards would make his most audacious acquisition in 1999 with the purchase of BP Amoco's Canadian crude oil assets for $1.6 billion in cash, scooping up the properties before other potential bidders could get organized. The deal boosted the oil output of Canadian Natural Resources and Penn West by 50 per cent.[37] Canadian Natural Resources' shares climbed to $36.50 and Penn West's shot up, too. Three-quarters of the acquired production was the company's Mic Mac oil sands properties north of Fort McMurray, which became the foundation of Canadian Natural Resources' vast Horizon operation. The deal "rang the bell," said one senior oilman. "If it's good enough for Murray Edwards, it's good enough for me."[38]

CHAPTER 5
BIG OIL AND THINK TANKS BESIEGE KYOTO

The day after the Kyoto Protocol was signed in 1997, Tom d'Aquino went on the attack. The agreement was unrealistic, he said, because it committed Canada to reducing greenhouse gas emissions by 6 per cent no later than 2012. A much longer lead time was needed. Asking companies to make major changes all at the same time would entail huge additional costs and result in important losses in economic output. He pointed out that each plant or technology's normal lifespan is fifteen, twenty or thirty years before it must be replaced or upgraded. "It is in businesses' interests to achieve higher energy efficiency and become more productive," he argued.[1]

But that must happen through voluntary measures. He reiterated this message at a presentation to the National Forum on Climate Change, a body established by Ottawa's National Roundtable on the Environment and the Economy to garner citizen opinion on the subject.[2] Kyoto was unrealistic, d'Aquino told the forum, but business was committed to addressing the issue in a responsible fashion.

Industry was already acting responsibly because it had improved energy efficiency on a per-unit-of-output basis. And many firms — over 600 — had signed up for Canada's voluntary emissions program. The overwhelming emphasis, BCNI said, should be on providing appropriate incentives to encourage businesses and consumers to voluntarily reduce their emissions. Market-based instruments such as emissions trading and joint implementation that are included in Kyoto are better than government regulation — command and control, he called it — because they have the advantage of achieving the same environmental improvement at lower cost, or so he claimed. Canada should not ratify Kyoto until the US does.

D'Aquino had to tread a fine line between the divergent views of his members. Following the denialist views of American parent ExxonMobil, Imperial Oil ran an extreme campaign against climate change. At the company's 1999 annual meeting, president Bob Peterson warned of higher prices, increased unemployment, lower investment, reduced competitiveness and adverse impacts in all regions of the country should Canada implement the Kyoto targets. Even though the company was well aware of global warming's reality, Imperial published a position paper in 1998 challenging the scientific findings, dismissing environmentalists as "alarmists" and contending that it made more sense for Canada to do nothing.[3] Royal Dutch Shell, parent of Shell Canada, on the other hand, was spending US$500 million to develop new energy sources. The Canadian subsidiary followed in the footsteps of its European parent, opening the door to accepting that global warming was real and had to be addressed.

The same split could be detected in the Canadian Association of Petroleum Producers. To prepare for the assault on Kyoto, CAPP brought in a semi-outsider as president, not beholden to either side, but still well situated within the industry. Pierre Alvarez had been deputy minister of Energy Mines and Petroleum Resources in the Northwest Territories government and CEO of the Northwest Territories Power Corporation. And he knew where the bodies were buried in struggles between Ottawa and Alberta. He had been Pat Carney's executive assistant during the years she was dismantling the National Energy Program.[4] Alvarez would lead the organization until 2008, the year after industry pivoted from denial and voluntary measures to carbon pricing and clean growth. Along with d'Aquino and Ken Smith, Alberta's deputy energy minister, Alvarez masterminded the industry's assault on Kyoto ratification.

They took nothing for granted in making the case that Canada could not afford to ratify Kyoto. BCNI vice-president John Dillon authored a report insisting "Canada should not consider ratification of the Kyoto Protocol until there is a much better understanding of the full implications for the country, the rules of the game internationally are clear, as well as actions to be taken by our trading partners and competitors."[5] The report's backers included Imperial Oil, the Coal Association of Canada and CAPP. Even Chrétien's energy and industry ministers expressed concern about the impact Kyoto could have on the Canadian economy.[6] An Industry Canada study said that Kyoto would effectively impose a carbon tax on energy producers, an unthinkable option at the time.

But Chrétien was unmoved. In a 2001 letter to the BCNI and Canadian Chamber of Commerce, he said he intended to have Kyoto ratified. Industry could reap economic rewards by developing clean-energy technologies, he replied. "I am of the view that Canada should strive to be first."[7] As the prime minister who made possible continued oil sands development through his government's generous tax subsidies, Chrétien should have had some explaining to do. But it was becoming clear that the prime minister, who had said he wasn't running again, was seeing Kyoto as a legacy issue. That Canada would have to take serious action to cut emissions didn't seem to be part of his thinking since it would happen after he left office.

A "MADE-IN-CANADA" SOLUTION

As the clock ticked into 2002 and toward ratification, Chrétien's bureaucrats changed their tune from doubt and voluntary measures to support for Kyoto. To boost Kyoto's prospects, they downplayed its impact. A leaked Natural Resources Canada report estimated that Kyoto would cost oil sands producers only 11–13 cents a barrel, or just 0.5 per cent of the current price. Pierre Alvarez didn't believe the numbers. And Murray Edwards, who was readying his Horizon oil sands mining and upgrading project for start-up, said his company estimated Kyoto would cost between 50 cents and $7 a barrel. He was considering building the CAD$4.2 billion upgrader in the US rather than Alberta because of the uncertainty over Kyoto.[8]

Business was taking an additional tack. Instead of just opposing Kyoto, it concocted the idea of a made-in-Canada solution to climate change. This idea turned into the "single largest political campaign intended to influence environmental policy" since the beginnings of such policy in the 1960s, comments political economist Douglas Macdonald.[9] The made-in-Canada frame was mentioned first by Bob Peterson, Imperial Oil's outgoing president, in March 2002. "I am of the view [the federal government is] going to rethink this and come to a conclusion that there's got to be a made-in-Canada solution," Peterson told business reporters in Calgary.[10] He made this comment a month after US president George W. Bush unveiled an industry-friendly plan that would barely slow emissions of carbon dioxide and other greenhouse gases, let alone cut them.

Two weeks after Peterson's news conference, the phrase popped up again, emanating purportedly from Alberta environment minister Lorne Taylor. A *National Post* headline writer wrote "Alberta offers Kyoto alternative: 'Made-in-Canada' solution to greenhouse gas emissions to be unveiled at energy summit." But what Taylor said was that the province was working on a "US–style plan to reduce greenhouse gases."[11] Alberta's plan, Taylor said, would be based on the same principles as Bush's recent climate-change policy. It emphasized financial incentives for developing technologies that reduce carbon emissions. So the made-in-Canada plan would be based on the made-in-the-US plan, which would not cut emissions, just slow down the rate of increase; allow voluntary, not mandatory, reporting of emissions; and provide tax credits for renewable energy and fuel-efficient cars. These credits looked good but were too small to have much impact. It was a conscious attempt to mislead Canadians into believing a credible alternative existed that could actually slow global warming.

Three weeks after Taylor's announcement, Alberta premier Ralph Klein addressed CAPP's annual dinner in Calgary. He said he would work with the oil industry and the United States to find a "North American solution" or a "made-in-Canada solution" to climate change, using the phrases interchangeably.[12]

NATIONAL PUBLIC RELATIONS LAYS ASTROTURF

Then two things happened. "US–style plan" and "North American solution" disappeared from the rhetoric, leaving only the "made-in-Canada solution." And National Public Relations, the Canadian affiliate of giant PR firm Burson-Marsteller, took over the campaign to defeat Kyoto ratification. The oil patch had panicked when Chrétien said that he hoped to ratify Kyoto by the end of the year. On September 26, National Public Relations launched the Canadian Coalition for Responsible Environmental Solutions, whose mission was to instill the phantom made-in-Canada and responsible solution in the public mind.[13]

The Canadian Coalition for Responsible Environmental Solutions was a textbook example of an "astroturf" organization and Burson-Marsteller wrote the book on such entities. (As we saw in Chapter 3, Burson-Marsteller ran the 1992 Business Council for Sustainable Development campaign to deter the Rio Conference from considering mandatory caps on emissions.) If the oil and auto industries opposed controls on global warming, they would be dismissed as special interests and ignored. But if they could bring in other business and community interests and created the impression of a grassroots campaign, they would have a better chance of winning. As Rick Hyndman, CAPP's senior policy adviser on climate change and yet another former Alberta deputy minister of Energy, later acknowledged, Canadians were unlikely to view CAPP, the voice of the oil patch, as a credible source.[14] Time for an astroturf coalition.

The coalition took great pains to point out that it represented a broad cross-section of Canadian industry, that it was made up of representatives of more than twenty-five business organizations, industry associations and consumer advocacy groups, and that the members of the coalition accounted for the vast majority of Canada's private sector jobs, investments, exports, training and research and development.[15] These were not just the fossil fuel guys protecting their turf, they were saying. But behind the seemingly wide representation was the fact that virtually all the money that went into organizational activities, such as expensive saturation TV ads in Ontario, estimated to cost CAD$225,000 a week, came from the deep pockets of petroleum producers.[16]

National Public Relations' goal was to undermine support for the prime minister by drumming up as much grassroots outrage as possible about Kyoto and directing it at Ontario MPs, who formed the bulk of Chrétien's majority in the House of Commons. (In the 2000 election, the Liberals swept up 100 Ontario seats but netted only 14 across the West.) A primary target was the seventy-member Liberal caucus of Ontario MPs with petrochemical, auto-parts or vehicle-assembly factories in their ridings. These operations could be affected by measures to meet the Kyoto goals.[17]

The Canadian Coalition for Responsible Environmental Solutions put further pressure on MPs with the release of a paper entitled "We Can Do Better: Achieving a Made In Canada Climate Change Plan." D'Aquino spoke directly to MPs:

> *Parliament has been given notice by the Prime Minister that a vote will take place on the ratification of the Kyoto Protocol. We are appealing to each and every Member to exercise your prerogatives fully, to ensure that your constituents understand what is at stake and that sound and responsible alternatives are considered prior to the matter being put to a vote.*[18]

The campaign was backed by a poll that made the front page of *The Globe and Mail* timed to coincide with the "We Can Do Better" paper. "Support for Kyoto plunges," the headline screamed.[19] The *Globe* reported a survey by Ipsos-Reid conducted on behalf of the Alberta government and released in early November, 2002, suggesting that 44 per cent of Canadians wanted the federal government to ratify Kyoto, while 45 per cent wanted Ottawa to withdraw from the accord and develop a made-in-Canada plan. A poll done a month earlier by the same firm found that 74 per cent of Canadians supported implementing the accord, with only 21 per cent opposed.

This dramatic, sudden collapse in support for Kyoto was disastrous news for environmentalists, who had fought for cuts to carbon dioxide emissions since the alarm about climate change was first raised in 1988. It was also bad news for Chrétien, who pledged his government would ratify the accord by the end of 2002, just two months away. How could he proceed when Canadians were turning against it?

It was good news for the anti-Kyoto government of Alberta. Yet what was to be expected, given that the province had commissioned the poll in the first place and timed its release to contribute to the responsible solutions campaign? But all was not as it seemed. As became quickly evident, the reason for the different result was due not to a collapse of support for Kyoto but to the addition in the second poll of a third option to withdraw from Kyoto and develop a made-in-Canada solution. The pollster, Ipsos-Reid, was criticized for including this option because it was not provided in its earlier survey. The results of the two surveys were not comparable and the *Globe* article was fraudulent, federal environment minister David Anderson charged.[20] Ipsos-Reid president Darrell Bricker defended his firm's use of the phrase which, he said, is legitimate and indicates there is an "emergent view . . . that Canadians are open to a 'made in Canada solution' that may be different from the Kyoto Protocol."[21]

Bricker's problem, though, was that a made-in-Canada solution to climate change didn't exist. He admitted as much in his defence of the poll question. "To suggest that the phrase is irrelevant to the current debate on the accord is to deny the importance of rhetoric in shaping public opinion. As social scientists, it is our job to test the impact of rhetoric." The *Globe's* own Hugh Winsor described the rhetoric succinctly: "Why should we subject ourselves to some foreign deal struck an ocean away when the alternative is a good old homegrown alternative. This is a no-brainer . . ."[22]

A few days after National Public Relations' saturation ads wrapped up, a second PR firm was brought in to do battle with the accord. Imperial Oil and Talisman Energy were dissatisfied with the effectiveness of the made-in-Canada campaign and hired the Association of Public-Safety Communications Officials Canada (APCO), the Canadian subsidiary of a lobbying firm based in Washington, DC, to organize a gathering of scientists "skeptical" of the agreement, as part of a last-ditch effort to derail ratification. APCO specialized in supporting rogue scientists who were financed by industry and purported to challenge established scientific thinking. It made its mark attacking epidemiological studies that implicated second-hand smoke in increased rates of lung cancer and heart disease in non-smokers and financed by Philip Morris and other tobacco companies.[23]

APCO brought to Ottawa a roster of climate-change deniers for a conference to reveal Kyoto's "science and technology fatal flaws."[24] But they had minimal impact on federal MPs and bureaucrats. They were too late. Chrétien issued a directive that the vote to ratify Kyoto would be regarded as a confidence motion and the government could fall if Liberal backbenchers voted against it. Despite their misgivings, some backbench Liberals voted for ratification, some stayed away, and it passed. APCO's last-minute conference and National Public Relations' expensive propaganda campaign did not shift public opinion enough to create the change desired by industry.[25]

SUCCESS BEHIND CLOSED DOORS

But if industry couldn't win in the court of public opinion, it could still flex its muscle behind closed doors. While TV ads from the Canadian Coalition for Responsible Environmental Solutions were lambasting the government, CAPP was secretly negotiating with its champion in government, Natural Resources Canada, to limit the impacts of Kyoto. A deal was not long in coming. Five days after ratification, Natural Resources Minister Herb Dhaliwal wrote to CAPP chair John Dielwart that "on the price of carbon credits, the government will ensure that . . . Canadian companies will be able to meet their emission reduction and responsibilities at a price no greater than CAD$15 a tonne . . ." It was a coup for the fossil fuel industry because projections suggested that a price on carbon of CAD$100 to $250 a tonne would be necessary to meet the standard established by Kyoto. Taxpayers, not industry, would have to spend massively on international credits or domestic subsidies.[26]

CAPP continued to wield its awesome negotiating power during the Paul Martin government's term, which followed Chrétien's from 2003 to 2006. Martin was seen as more amenable than Jean Chrétien to fossil fuel industry persuasion. He certainly came from the business side of the Canadian polity. He started out as executive assistant to none other than Maurice Strong, who at the time was CEO of Power Corporation. Martin was appointed to the board of a Power Corporation subsidiary Canada Steamship Lines, a Montreal-based shipping company that operated mainly on the Great Lakes. He eventually became president and with the help of deep-pocketed partners, purchased the company for CAD$195 million in 1981 (about $540 million in 2018 dollars).

First elected to Parliament in 1988, he was appointed minister of Finance in 1993 after Chrétien won a majority government. Martin's time as finance minister was notable for his 1995 budget that slashed social spending. He ran for the Liberal leadership in 2002 and even though he was assured of victory, the funds kept rolling in. He ended up pulling in a record CAD$12.2 million, almost entirely in large donations.

Were corporations expecting future benefits once he was in power?[27] Murray Edwards would know. He was one of Martin's chief fundraisers. Edwards had been a loyal federal Liberal since the 1980s, a lonely voice in Calgary, especially during the days of the NEP. Edwards was especially loyal to Martin. After Martin's 2004 election victory, Edwards and his wife Heather Edwards hosted Martin at a dinner at their Calgary house to thank party faithful for donating to the last federal campaign. And Martin showed that they were in sync big-time. In 2005 his government released another climate change plan, revealing that the total greenhouse gas emissions reduction required from large industrial polluters — oil sands plants would be near the top of the list — were cut by a further 45 per cent beyond CAPP's deal with Natural Resources Canada.[28] "Business may have lost the Kyoto battle, but . . . was still winning the climate policy war," Douglas Macdonald observes.[29] By the time Martin's government was defeated in 2006, Canada was almost 30 per cent above its Kyoto target, compelling evidence of a successful defensive corporate lobbying strategy.

FRASER INSTITUTE'S DENIAL EFFORTS

If the lead hands on undermining Kyoto were CAPP and BCNI, the think tank wing was not letting up, applying the denialist tactic and then drawing other arrows from its quiver. The Fraser Institute was virtually silent on climate change until Kyoto was about to be signed in December 1997. American think tanks were churning out a torrent of material and it was inexpensive for the Canadian organization to recycle their publications.

The first product was a book of incendiary articles written by prominent American deniers, compiled by the institute's environmental economist, Laura Jones, and published in 1997, titled *Global Warming: The Science and the Politics*. Jones, an economics graduate from Simon Fraser University, used a major theme from the denial literature to frame the discussion: ". . . global warming is not a settled issue. The scientists [contributing to the book] raise important questions about the validity of the hysteria over global warming. It is time to revisit these questions."[30]

The institute used a formula it had perfected for the early Internet times: release a book, hold a one-day conference to discuss the book, seek print, electronic and digital media coverage. The book goes into libraries and is sold in bookstores; successful media coverage extends the book's reach. Jones introduced the conference with an op-ed in the *Vancouver Sun* critiquing the Intergovernmental Panel on Climate Change's use of computer models, a common theme for denialists.[31] The conference received positive coverage from libertarian commentators such as Terence Corcoran with the *The Globe and Mail* at the time — "there are fewer hurricanes today than ever before" — and Lorne Gunter in the *Edmonton Journal* — "global warming is due to solar cycles."

But the real payoff was a lengthy survey by science reporter Margaret Munro, also in the *Vancouver Sun*. Munro gave equal weight to, on the one hand, the vast majority of climate scientists and to the handful of denialists whose work was supported by the Fraser Institute and its corporate backers on the other.[32] This false equivalency had been established in media coverage and was a major triumph for denialists.

Jones released two more denialist publications over the next few years. The first was a 2001 pamphlet, *Global Warming: A Guide to the Science*, co-authored by Willie Soon and Sallie Baliunas of the George C. Marshall Institute, whose main point is that if global warming is occurring, it's because of solar activity and not greenhouse gas emissions. Jones also edited a tract titled *Facts, Not Fear: Teaching Children About the Environment*. This publication was aimed at teachers and school children, with an introduction by noted Canadian denier Patrick Moore.[33] The book was recycled from an American version published by the Alabama Family Alliance, a neoliberal think tank established in the late 1980s. A chapter on climate change quotes many of the same deniers as the institute's first publication. It tells parents how to talk to children about global warming:

> *Child's question: Is the world getting hotter?*
> *Answer: No one really knows.*
> *Question: Are human activities causing global warming?*
> *Answer: Perhaps . . . but the increase in warmth may be very small.*
> *Question: Is carbon dioxide harmful?*
> *Answer: No . . . it is a beneficial part of the atmosphere.*

In the year before Kyoto ratification by Parliament in 2002, Jones moved on to become chief lobbyist for the Canadian Federation of Independent Business in British Columbia, where she carried on her anti-regulation work. Her position at the institute was filled by Kenneth Green, who had formerly been director of the environmental health and safety program at the Big-Oil funded Reason Foundation, a Los Angeles–based neoliberal think tank that specialized in fighting environmental regulation, declaring it an affront to liberty.

Reason's anti–global warming propaganda is based on the assertion that "climate change has been a natural phenomena [sic] throughout the course of time." Do man's activities affect climate? "That is still open to debate," Reason's website said in 2004. "Some would say none at all while others will say that it is dramatic."[34]

Green promoted the same fossil fuel interests during his three years at the Fraser. He co-ordinated an issue of *Fraser Forum*, the institute's monthly publication, featuring six articles denying climate change.[35] Green's contribution, "Kyoto Krazy," reiterated the denial talking point that the science of global warming was still unsettled.

RESISTING KYOTO RATIFICATION

The lead-up to Kyoto ratification was a busy time for Green and the institute. It ran its own program to undermine the accord, intersecting with the National Public Relations campaign at various points. The first volley was a Fraser Institute-sponsored dinner in November 2002 to honour former Ontario Premier Mike Harris, who had been appointed a senior fellow at the institute. Harris compared Kyoto with the hated National Energy Program in a speech to 300 or so oil and corporate executives in a downtown Calgary ballroom.[36]

Harris's speech was upstaged a week later by Ralph Klein at a Fraser Institute–sponsored luncheon in Vancouver. Klein received an Adam Smith tie and an award for his "unrelenting commitment to fiscally responsible government."[37] He unleashed a blistering attack on the Chrétien government for implementing Kyoto and destroying the economy. Oddly, the Alberta premier's speech in west coast Vancouver received media coverage mostly in small-town Ontario newspapers, a target market for National

Public Relations' anti-Kyoto campaign. It was driven by a phrase Klein used to describe Kyoto — "a goofily concocted theory" — that fitted comfortably into newspaper headlines.

Kenneth Green chimed in with an op-ed claiming that the scientific consensus on global warming is more public relations than reality. It embodies the "simplistic 'turn-back-the-tide idea' that the only solution to global warming is to reduce greenhouse gas emissions." There is a second solution, he wrote, the made-in-Canada plan, referring to the Klein-National Public Relations fabrication.[38] Green billed himself as an "official reviewer of the most recent UN report on the science of climate change." That doesn't mean he was asked by the Intergovernmental Panel on Climate Change to review material. As *DeSmog Canada* explained, all it meant is that he asked to see the draft report and signed an agreement not to publicly comment on the draft.[39]

The institute next released a poll of large manufacturing companies in Ontario revealing that 62 per cent of respondents opposed ratification before the end of the year. And by a 12:1 margin, respondents said a made-in-Canada program was important.[40] Next the institute hosted a lunch in Vancouver for the authors of *Taken By Storm: The Troubled Science, Policy and Politics of Global Warming*, which was put out not by the think tank but by a Toronto publisher, Key Porter Books. The following day, they were off to Calgary for a Fraser Institute-sponsored book launch, which was reported as a newsworthy event in the *Calgary Herald*. Co-author Ross McKitrick was a University of Guelph environmental economist and Fraser Institute senior fellow and Chris Essex was a professor of applied mathematics at the University of Western Ontario. Their theme was classical denialism: fears about global warming and the alleged breakdown in the relationship between science and policy have led to a crisis that in all probability does not exist. Therefore, Kyoto is senseless policy.[41]

Green ended the campaign the day Kyoto was ratified with an op-ed comparing Kyoto with a lump of coal for Christmas (an interesting choice of symbol with its high carbon content). He reviewed the denial dossier: the threat of global warming was overstated by fear mongers like the David Suzuki Foundation; implementing Kyoto will be a waste of effort, will cost the average family CAD$3,300 a year, and will lead to considerably poorer people.[42] Finally, along with two other deniers, he wrote a report shortly after Kyoto ratification titled *The Science Isn't Settled*. Only one of the three, Steven Schroeder of Texas A&M University, had any training or expertise in climate science, but this handicap didn't seem to deter them. They made the claim that ratification of Kyoto relied on a simplistic computer model based on an inadequate database that omitted key economic factors and produced incorrect outputs.[43] (Their expertise in computer models was also limited.)

Green left the Fraser Institute in 2005 to become executive director of the Washington, D.C.–based Environmental Literacy Council, which provided industry-friendly learning materials for schoolchildren. This organization received funding from many of the same sources as the Reason Foundation: the American Petroleum Institute, ExxonMobil, the Sarah Scaife Foundation and the Charles G. Koch Foundation.[44] Green also became a visiting fellow at the American Enterprise Institute, where ExxonMobil CEO and denier Lee Raymond was vice-chair. Just before the Intergovernmental Panel on Climate Change's Fourth Summary Report was due to be released in February 2007, Green authored a letter on American Enterprise Institute letterhead offering US$10,000 each to scientists and economists for articles disputing the panel's findings.[45] It's not known how many scientists took the money.

The Fraser Institute also responded to the Intergovernmental Panel on Climate Change's report with its own study, designed to look like an official panel document. The "Independent Summary for Policymakers" was edited by Ross McKitrick, Green's pragmatic and scientific researcher, and written by ten prominent deniers, eight of whom including McKitrick, had been presenters at Heartland Institute denial conferences. The deceptive nature of the report could be seen in its conclusion, which began with this sentence: "The following concluding statement is not in the Fourth Assessment Report but was agreed upon by the . . . writers based on their review of the current evidence."[46]

This is deceptive because why would anyone expect a statement by a group of individuals with no connection to the Intergovernmental Panel on Climate Change — in fact openly hostile to that organization — to be included in its report? The summary's conclusion maintained denialist orthodoxy: ". . . there will remain an unavoidable element of uncertainty as to the extent that humans are contributing to future climate change, and indeed whether or not such a change is a good or bad thing."

THE TOBACCO PAPERS

Behind the Fraser Institute's anti–global warming, pro-industry work laid the support of most major oil, gas, petrochemical and coal companies in Canada. At least forty companies in the oil and gas production, exploration and transmission industries have been members.[47]

What did they get for their support? What might their money have bought? Former executive director Michael Walker has long claimed that institute members and directors could not tell the staff what to research, let alone dictate research findings. According to this view of the think tank's mission, oil and gas executives from Calgary could not tell institute staff to produce studies proving that global warming is not happening so we don't need to cut production. The staff undertook these studies on their own because that's what the research indicates. Or so one would have concluded from Walker's denials.

Walker's claim might be credible if we didn't have the Tobacco Papers — a series of letters Walker and his chief fundraiser wrote to a big tobacco company in 2000 asking for financial support for research that would cast doubt on the link between second-hand smoke and cancer. This rare glimpse into the role of corporate funding in shaping Fraser Institute research was obtained as a result of the 1998 Master Settlement Agreement between forty-six US state attorneys general and Big Tobacco. A condition of the agreement was that the tobacco companies had to make public and post on dedicated websites every document used in the discovery phase of legal actions brought by the states against the tobacco industry for Medicaid costs associated with smoking-related diseases. Documents created until 2010 were to be included. This trove of documents revealed at least one way that think tank funding and research intersected: The organization came up with a research program and then sought out corporations that would benefit from the research to finance the work.[48]

More than 80 million pages of documents were posted. They included letters written by Walker and Sherry Stein, Fraser Institute's fundraiser as of 2018, to the British American Tobacco Co., the world's second-largest tobacco company and owner of Imperial Tobacco (Imasco), which controlled 70 per cent of the Canadian market. The letters revealed that the Fraser Institute had set up a social affairs centre to promote free-market solutions to social policy problems like poverty, drug use, smoking and gun control.[49] Its aim was to demonstrate, through statistical legerdemain, that competitive markets provide a "better, more efficient [and] fairer basis for meeting popular aspirations" than "government control of social policy."[50] The institute's 1997 annual report states that the "majority of [its] revenues are derived from the donations of its members, and from research foundations. Revenue is also generated from the sale of publications, from Institute events, and from interest on invested endowment funds."[51] Somehow the institute forgot to mention the revenues raised from Big Tobacco. According to Stein's letter, tobacco company Rothman's International was providing $50,000 a year for this work and Philip Morris, "generous support."

What did Philip Morris expect for its generous support? An August 2000 PowerPoint presentation to Philip Morris International Corporate Affairs by the company's external affairs department states the mission of Corporate Affairs is "To respond to & shape a political, regulatory & attitudinal environment that permits the company to achieve its business objectives." The Fraser Institute is listed in the presentation as a non-US public policy organization that Philip Morris can utilize for "partnership/programming" in accomplishing the mission.[52]

With this funding, the Fraser published a book by two tobacco industry lobbyists titled *Passive Smoke: The EPA's Betrayal of Science and Policy*, and held two day-long conferences in Ottawa near the nation's lawmakers. This package of initiatives was timed to coincide with bylaws being considered by municipalities across the country to regulate smoking in public places. A bylaw banning smoking in all indoor public places had just come into effect in Victoria, B.C.

The Fraser's book argued that these bylaws were ill considered because the link between second-hand smoke and lung cancer had not been proven, a claim that was patently untrue.[53] The book attacked the landmark 1993 decision of the US Environmental Protection Agency that declared second-hand smoke to be a carcinogen. The Ottawa conferences worked in tandem. The first, titled "Junk science . . . junk policy," attacked the need for regulation of any kind; the second, with the motto "butt out," attacked the regulation of smoking. Neither the book nor the conferences mentioned tobacco industry funding.

At the end of 1999, Rothman's, Canada's number-two tobacco manufacturer, was bought by British American Tobacco (BAT) for CAD$11.5 billion, and the Fraser Institute lost this source of funding during its most prolific phase of pro-industry messaging. It quickly commenced a campaign to replace and add to the money. The institute already had one advantage in courting the funding: Brian Levitt, the CEO of BAT Canadian subsidiary Imasco, was on the institute board. The campaign included meetings in London, England and dinners in Toronto. A first letter, written by Fraser fundraiser Sherry Stein to BAT chair Martin Broughton, asked him to take over Rothman's funding commitment for the social affairs centre and to consider a new initiative for a risk and regulation centre. Stein asked for CAD$50,000 a year for each. BAT funding for this new centre would help the Fraser "provide the factual information that will seriously counter the risk activists and their misleading and misguided propaganda," Stein wrote.[54]

Later, in 2000, Laura Jones, who was also managing the institute's global warming denial campaign, thanked Adrian Payne, BAT's international scientific affairs manager, for "the most enjoyable dinner last week in Toronto."[55] Then Michael Walker reiterated these requests in a pitch to Payne. Walker focused on the new centre for studies in risk and regulation. He railed against the "agitators for a 'zero-risk' society [who] have become increasingly successful in advancing their cause, often basing their case on exaggerated junk science scares." The targets of these nasty agitators were environmental quality, second-hand smoke, pesticides and genetically modified foods. With BAT financial assistance, the Fraser Institute would set the record straight.[56] BAT would not be alone in supporting the new centre, Walker reassured Payne. The institute had already contacted virtually every major company in the industry:

[We have] met with a number of your colleagues in the industry to discuss this proposal and all are on side and have implied that they will support the Centre with comparable contributions. The companies they represent are Imperial Tobacco Company Ltd., JTI Macdonald Corporation, and Rothman's Benson & Hedges Inc. We have begun discussions as well with Philip Morris International Inc., and Brown and Williamson Tobacco in the US. Others we will approach for support are in the food, biotechnology, and chemical industries.[57]

Stein presented three proposals for BAT's support. It could contribute CAD$30,000 for the launch of the centre featuring guest speaker John Stossel, a well-known television personality and anti-regulation zealot; $42,000 to distribute an anti-regulation book called *Safe Enough?*; or $48,000 for a project that would show regulation was too costly to be effective or all three options.[58] The documents don't indicate which, if any, of these projects BAT did support, but they all took place.

Just as the institute was expanding its tobacco programming, Big Tobacco cut its funding for think tanks and front groups that were manufacturing doubt about tobacco's harmful effects in the mature North American and European markets. The money was directed instead to similar organizations operating in the high-growth markets of Eastern Europe, Russia and Asia. Fraser discontinued its smoking campaign, but didn't lack for funding. Other industries with serious health and environmental impacts from the use of their products stepped up to the plate to continue supporting front-group work. And Big Oil money was already flowing into its coffers. However, unlike the tobacco industry, agreements between these funders and the neoliberal think tanks they support remain locked away.

CHAPTER 6
CARBON PRICING: ECONOMISTS VS. THE PUBLIC INTEREST

In a 2016 *Globe and Mail* opinion piece, economists from the Ecofiscal Commission, a prominent Canadian pro-carbon-pricing advocacy group, said this: "Either economy-wide carbon tax or cap-and-trade systems reduce greenhouse gas emissions at a lower overall cost than 'command-and-control' government regulations."[1] They provide no evidence of the superior cost effectiveness of carbon pricing. They don't have to. It is common sense — a "fact" accepted by almost all economists, corporate executives, media commentators, environmentalists and politicians, at least those who believe global warming is occurring. But it is a different kind of fact than, say, that Ottawa is the capital of Canada. It is physically impossible to run a real-world experiment comparing the effects of carbon pricing and direct government regulation in the public interest — the actual name for command-and-control — for the same jurisdiction over the same time period. You can't rewind and do over again. This fact wasn't observed in nature or society; it was constructed by economists over decades. And it was just the fact the fossil fuel industry was looking for, because it removed the threats of direct government regulation and imposed technological standards.

Carbon pricing — both carbon tax and cap-and-trade — can be traced to the work of British economist Arthur Pigou (1877–1959) and his 1920 book, *The Economics of Welfare*. Pigou, who taught at the University of Cambridge for most of his career, was concerned about the role government policy could play in solving the problem of poverty and increasing national well-being. Key to Pigou's impact on climate change policy was the concept of an externality: a cost imposed or a benefit conferred on others that is not taken into account by the person taking the action. Pollution is an example of a negative externality; education a positive one. The person creating a negative externality such as pollution, if not checked, will engage in too much activity that generates the externality, and will thus upset the general equilibrium in the economy. Someone creating a positive externality such as educating herself perhaps will not invest enough in her education because she does not perceive how valuable her education may be to society. The negative activity is oversupplied while the positive one is undersupplied. Pigou saw externalities— both positive and negative — as impediments to progress that justified government intervention.[2]

To discourage an activity that causes a negative externality and to encourage an activity that causes a positive one, Pigou proposed a system of taxes and subsidies. A Pigouvian tax is a tax on goods with negative externalities equal to the costs those goods imposed on society. A Pigouvian subsidy has the opposite effect.

The idea that pollution generates a social cost that should be addressed by the central government marks the beginning of modern environmental economics that so dominates current global warming thinking today, including the preoccupation with carbon taxes. In every case, though, carbon taxes that have been applied in the past decade are not Pigouvian because they do not equal the cost imposed on society by the pollution. They would have to be in the range of CAD$150–$300 per tonne of carbon dioxide equivalent — and perhaps much more — to achieve that goal.

Such a drastic charge would cost elected politicians their jobs and likely ruin the capitalist system. So two economists, William Baumol and Wallace Oates, in 1971 proposed that government could determine an acceptable level of pollution and set the tax at a level that would achieve that standard. It was a distortion of Pigouvian theory, but it gave economists a toehold in environmental pollutions issues.[3] And subsidies, say to industries producing zero-carbon emitting technologies, seemed to be held in low regard by many economists. But the desire for revenue neutrality is another violation of Pigouvian taxation since positive externalities, such as better public transit or energy-efficient buildings, should be subsidized to achieve a more efficient market outcome.

COASE'S CRITICISM OF PIGOU

Pigou's proposal that a central authority should impose taxes and confer benefits to internalize externalities within the economic system was acceptable to mainstream economists because it was a case in which government intervention could improve the functioning of the economy. It came under attack, though, from the neoliberals associated with the Mont Pèlerin Society, some of whom taught in the economics department at the University of Chicago and at the London School of Economics. They rejected Pigou's approach for its naïve "benevolent despot" assumption, so named because government could never know enough to make the best decisions; only the market was capable of making them through its billions of daily transactions.[4] As Chapter 2 explains, during the 1960s, neoliberal economists were developing theories that privileged property rights over social rights, applying market principles to vast swaths of social and political life.

Ronald Coase (1910–2013) brought market thinking to environmental issues, bypassing Pigou's vision of a direct government role. Coase was a British economist who worked at the University of Chicago for most of his career and won the Swedish National Bank's Prize in Economic Sciences in Memory of Alfred Nobel (incorrectly called the Nobel Prize in Economics to give the discipline the same cachet as physics, chemistry, medicine, literature and peace) for his 1960 book, *The Problem of Social Cost*.

The key point the book makes, and the one that made Coase a hero to neoliberals, is his explanation of how market economics, and not the "heavy hand" of government, as they saw Pigou's approach, can address pollution problems. Taxes and regulations are unnecessary, Coase held, because in some cases polluters and those harmed by pollution can engage in private negotiations to determine the appropriate compensation. All that is required is a system of clearly defined property rights. Private citizens or firms "can negotiate a mutually beneficial, socially desirable solution as long as there are no costs associated with the negotiation process."[5]

The doctrine has not been commonly applied because Coase's fundamental assumption that negotiation between parties is costless is rarely correct.[6] Citizens might face considerable costs to gather scientific and legal information, prepare legal documents and enforce the company's compliance with a contract between the parties. These drawbacks were ignored since Coase's work was in the vanguard of the counterattack on the burgeoning environmental regulation movement of the 1960s. Coase's formulation soon led to the idea of creating property rights for noxious substances instead of taxing, banning or regulating them, and this idea influenced economic thinking for decades. Pigou's approach would have to wait until after the turn of the century before it was seen as the alternative to regulation.

Six years after Coase published his book, University of Wisconsin graduate student Thomas Crocker proposed capping emissions of pollutants and then letting firms buy and sell permits that allowed them to pollute within the cap.[7] A second economist, a Canadian named John Dales, proposed a similar approach for Canadian farmers who were polluting lakes and streams. In his 1968 book, *Pollution, Property and Prices*,[8] Dales explicitly acknowledges Coase's influence on his thinking.[9] Several years later, a third economist, David Montgomery of the California Institute of Technology, "proved" that Crocker and Dales' concept could minimize the overall cost of achieving a given emissions reduction.[10] Through these works, pollution was being commodified.

CONTROLLING ACID RAIN

Emissions trading was still a theory when the US Congress passed, and Richard Nixon signed into law, the *Clean Air Act* in 1970, two years after Dales published his book. According to American business historian David Vogel, the legislation was part of "the most far-reaching set of pollution-control requirements ever imposed on industry."[11] It required the newly created Environmental Protection Agency (EPA) to set national air-quality standards for six pollutants that had been determined to have environmental impacts. They included sulphur dioxide, carbon monoxide and nitrogen oxide. (Carbon dioxide was not yet considered a pollutant.)

The EPA was also required to establish detailed timetables and deadlines for enforcing its various provisions. The act's uncompromising approach to quality standards resulted from its being based on public health considerations and not on economic impact or technical feasibility. The EPA would consult with doctors and scientists, not economists. If a pollutant threatened to sicken or kill people, then it must be reduced in impact or abandoned; never mind how much it might harm an industry or how expensive technical fixes might be.

The legislation even included a provision authorizing citizens to file lawsuits in federal court to force compliance by the EPA. That's why Vogel called the *Clean Air Act* "one of the strictest, most controversial and bitterly fought pieces of regulatory legislation enacted by the federal government."[12] Canada's *Clean Air Act*, passed the following year (1971), was based on similar health concerns regarding the "detrimental relationships between humans and airborne contaminants."[13] Early studies were undertaken by the Department of National Health and Welfare. The Department of the Environment was created in tandem with the act. It came to regulate mercury, vinyl chloride, asbestos and other pollutants.

There was nothing ineffective about the American act. It forced industry to move to cleaner methods of production. Sulphur dioxide emissions were reduced by 25 per cent and particulate emissions by 14 per cent within six years.[14] With its stringent standards, the law was loathed by industry, which went all out to resist any further mandatory regulations. Industry's ultimate triumph was to affect a transition from protecting public health in the 1970 act to promoting market efficiency in the next major iteration of the act in 1990.

Industry succeeded because of the indispensable help it received from a segment of the economics profession. It was an example of what Italian Marxist theoretician Antonio Gramsci (1891–1937) labelled "organic intellectuals."[15] All people are intellectuals, he wrote, but not all people have the function of intellectuals in society. These are teachers, clergy, professors, managers, lawyers, economists and others who have developed organically alongside the ruling class and function for its benefit. They are "produced by the educational system to perform a function for the dominant social group in society. It is through this group that the ruling class maintains its hegemony over the rest of society," explains Barry Burke, who teaches working-class history in the UK. By hegemony, Gramsci meant "the permeation throughout society of an entire system of values, attitudes, beliefs and morality that has the effect of supporting the status quo in power relations."[16]

ECONOMISTS ELBOW THEIR WAY IN

The 1970 US *Clean Air Act*, with its focus on health and its reliance on scientists and medical doctors, was perceived by economists to have shut them out from the burgeoning field of environmental policy implementation.[17] Inspired by the nascent neoliberalism in the air, as emblemized by the Coase-Crocker-Dales approach — the new hegemony in the making — it was not long before the EPA adopted an economic approach to enforcing the act and hired young "entrepreneurial" economists as staff members in its Office of Planning and Evaluation. Economic commentators were attacking the clean air regulations as being ineffective and politically unrealistic. They urged policy makers to replace the act's technology-based standards with market incentives such as charges, taxes or marketable permits.[18]

Careful studies of the EPA suggest that economists were not merely studying already existing markets because these markets didn't yet exist. Instead, insists Donald Mackenzie, a sociologist at the University of Edinburgh, economics was helping "bring a new market into existence."[19] Economics was not being descriptive or analytical, but had become a player, creating the phenomenon it studies and describes. These economists went beyond the traditional role of studying the economy to a new one in which they promote markets, thus creating the economics profession's own version of what an economy should do.

The first move toward markets was the EPA's 1979 "bubble" policy, which allowed some polluting facilities to avoid costly charges at their plants to meet air quality standards. Instead of enforcing limits on each source of pollution, such as a coal-fired power plant, the policy allowed an overall target for each pollutant type. It was as if the total facility was covered by a bubble. By making improvements in some parts of the facility, a company could continue to pollute in other parts. Bubbles saved money for polluters, but meant that optimal progress toward achieving ambient air quality standards would never be met.[20] The EPA also allowed companies to offset additional emissions from a new facility by reducing emissions elsewhere in the same airshed.[21]

Environmental groups repeatedly raised legal challenges to the policy, but it was ultimately upheld by the Reagan Supreme Court. This line of work by economic theorists "pulled the nascent policy scheme out from the shadow of the [direct regulation] regime and highlighted it as a first instance of a new policy instrument in practice, a proof of the principle that emission reduction obligations could be traded," writes Jan-Peter Voss of the Technical University of Berlin.[22] Spurred on by the Coase theorem, economists had succeeded in creating a new reality. Now they could devote themselves to applying it. They were so successful that their creation was embedded in the Kyoto Protocol and became the prototype for dozens of emission-trading schemes, including Ontario Premier Kathleen Wynne's thirty years later.

FROM PUBLIC INTEREST TO COMMAND AND CONTROL

To transform the purpose of the *Clean Air Act* from protecting health to promoting efficient market transactions, business and its economist allies needed to accomplish three things: frame the act as "command and control" regulation evoking images of Soviet Union-style authoritarianism; distinguish command-and-control from incentive-based mechanisms; and prove that incentives are more efficient than command and control. Prevailing wisdom would have to be shifted from regulation as the best way to protect the public interest to market-based transactions as the most efficient way to boost the economy. This process could be seen as a struggle for turf. Engineers, scientists and lawyers played a central role in traditional regulation practices. If economists were to make incursions into the field, one strategy was to demean the traditional approach. And it worked, as lawyers and scientists came to fear these forays into their field. But they never fought back.

The idea of the "public interest" is meant to counter the self-interest of powerful individuals, corporations and well-organized groups. The unorganized mass of society cannot be expected to look after the public interest in the same way. So it falls to the state to defend the public interest.[23] The state creates regulations to control corporate behaviour with undesirable health, safety, consumer or environmental consequences. Typical regulations limit the prices of a natural monopoly such as a telephone system, impose safety measures on airlines and license doctors to prevent charlatans from treating patients.[24]

During the 1960s, public interest regulation was considered the most effective means of controlling negative corporate actions. Rachel Carson's *Silent Spring* accused the pesticide DDT of poisoning bird populations and threatening human health. Lawyers and scientists combined skills to promote a ban on the chemical, which was accomplished in 1972. The burgeoning environmental movement had also pushed through the *Clean Air Act* and the 1972 *Clean Water Act*. These, along with the DDT ban, were among the greatest achievements of American public interest regulation.

Was public interest regulation nothing more than command and control?[25] In 1977, economist Charles Schultze, chair of US President Jimmy Carter's Council of Economic Advisors, wrote a book, *The Private Use of Public Interest*, which uses command and control several times. In a *New York Times* article titled "There must be something besides rules, rules and more rules," Schultze reiterated the book's theme that "the current 'command-and-control' approach to social goals, which establishes specific standards to be met and polices compliance with each standard, is not only inefficient 'but productive of far more intrusive government than is necessary.'"[26] Schultze complained that government was run by lawyers, who "concentrate on rights and duties as well as equity, when it should be run by economists who focus on efficiency and market considerations."

There is some logic, it's true, in applying the term to regulation. Command refers to the standards or targets set through government legislation that must be complied with by the members of the target industry. Control refers to the negative sanctions such as fines, penalties or prosecutions that can be utilized to bring a non-complying entity into accord with the regulations. Nonetheless, its use during the heyday of the Cold War sent an unambiguous message — beware the heavy hand of government. And why use litigiously minded lawyers when market-friendly economists were ready to help?

Schultze was not associated with the neoliberals in the Mont Pèlerin Society or in the University of Chicago economics department, but with the centrist Brookings Institution, indicating how far neoliberalism was infiltrating the liberal establishment during the 1970s. He was a mainstream Democrat who had served as budget director for President Lyndon B. Johnson.

Schultze became the standard-bearer for deregulation, or, as it was called in respectable circles, regulatory reform. The fact Schultze was appointed by the Democratic president Jimmy Carter lends support for those, like David Harvey, who argue that neoliberalism started in the United States under Carter, not Ronald Reagan.[27] Two months after taking office, with Schultze as his chief economic lieutenant, Carter issued executive order 12044, "Improving Government Regulations."[28] This order "directed regulatory agencies to find ways to achieve their goals with reduced burden on the private sector."[29] Carter, guided by market enthusiast Schultze, appointed market enthusiast Douglas Costle to head the EPA and move markets forward.

Command and control stuck, as corporations, economists and politicians piled on. Carter's own memo lauding the success of his 1977 order referred to his government's "traditional approach of rigid, detailed 'command-and-control' regulation."[30] It became so firmly entrenched in creating a negative frame for government regulation that no one even noticed. With a nod to Coase-Crocker-Dales, the public interest has disappeared and lower overall cost has become the measuring stick for controlling pollution.

Schultze also helped along a more widely used epithet for public interest regulation — red tape. In his 1977 book, Schultze claimed:

> [T]he public has become disenfranchised with the ability of government . . . to function effectively . . . The rash of new regulatory mechanisms established in recent years — for pollution control, energy conservation . . . — has generated a backlash of resentment against excessive red tape and bureaucratic control.[31]

The term red tape has been around for centuries. It was first used to refer to the red ribbon wrapped around important documents. By the nineteenth century, red tape referred simply to rules. Schultze tied it to what he claimed was excessive regulation. However, a leading expert on regulation doesn't think red tape is related to the number of regulations. Barry Bozeman of Arizona State University defines red tape as "rules, regulations, and procedures that have a compliance burden but do not achieve the functional objective of the rule."[32] Rules that are effective are not red tape. This crucial distinction was lost as the Reagan administration settled into power with its full-court-press anti-regulatory agenda.

TRADEABLE PERMITS BECOME A REALITY

With the negative framing of direct regulation taking hold, tradeable permits were embraced as the way forward, leaping from economic theory to practical policy in 1988 when George H.W. Bush won the US presidential election, based partly on his promise to be the "environmental president." The Coase-Crocker-Dales approach to environmental problems looked promising to free-market Republicans who had faced an existential dilemma during the Ronald Reagan presidency.

Thanks to the undermining of the 1970 act by the bubble concept and the use of offsets, American coal-fired power plants were still emitting sulphur dioxide, which was returning to earth as acid rain, damaging lakes, forests and buildings in eastern Canada and the northeastern United States. Still, the prevailing environmental view followed the regulatory regime of the 1970 *Clean Air Act*. Utilities responsible for emitting the pollutant must be required to install scrubbers in their smokestacks to remove sulphur dioxide from power plant exhausts. The 1970 act was still working well. In 1990, aggregate emissions of the six leading pollutants were down 33 per cent since 1970.[33]

But power companies had been resisting almost since day one. They continued to insist that the cost of installing scrubbers would bankrupt them.[34] Republicans couldn't let their coal company backers down. At the same time, they were faced with the harsh reality of environmental damage. They were at an impasse. Seventy acid rain bills were put forward by Congress; none made it into law. What Republicans needed was a free-market solution to acid rain. Then along came Fred Krupp of the Environmental Defense Fund (EDF) and he had the solution.

The EDF was a traditional scientist- and lawyer-run "sue the bastards"

environmental advocacy organization formed five years after Rachel
Carson's *Silent Spring*. EDF successfully fought to ban DDT, first in New
York state and then nationally in 1972. It helped pass the *Safe Drinking
Water Act* in 1974 and phase out lead from gasoline in 1985. But the foll-
owing year, it hired lawyer Krupp as its executive director, and he set off in
a new direction.

By this time, neoliberalism had swept through the economics profession
and was infiltrating many environmental organizations. And from a pragmatic point of view, advocating the use of markets to solve environmental
problems was likely to attract significant corporate funding, an additional
bonus for a pro-industry environmental organization. Environmental economics would be "his horse to ride," Krupp said, and he rode it well.[35]
Krupp started replacing EDF's lawyers with economists. In a 1986 op-ed
in the *Wall Street Journal*, he argued that it was no longer sufficient for
environmentalists to oppose pollution. They needed to carefully consider
the impact of regulation on industry, taking into account growth, jobs and
shareholder interests. He coined the term "third stage environmentalists,"
who don't just oppose the problem but also find a way to meet the need.[36]
It didn't take long for his market-driven approach to attract "Wall Street
people, who would become the most important trustees and benefactors
for EDF over the years," writes Eric Pooley in his chronicle of the American
climate wars.[37]

The article also caught the eye of C. Boyden Gray, chief counsel for Bush
the elder, then Reagan's vice president, who was positioning himself to
run for president. Gray saw how Krupp's approach could help Republicans
solve their dilemma and benefit Bush's ambitions. He invited Krupp to the
White House for lunch. Meanwhile Krupp asked Republican senator John
Heinz (Pennsylvania) and Democratic senator Tim Wirth (Colorado) to
assemble a coalition of players from both parties, and from industry and
environmental NGOs, to push for a market-based approach to environmental governance.[38] A former EDF staff economist was recruited to lead
the effort, which resulted in a report titled "Project 88: Harnessing market
forces to protect the environment." The report included a proposal for a
market-based approach to reducing acid rain.[39]

The report sucked all the oxygen out of the policy area, leaving a market solution as the sole candidate standing. Just as Bush was readying to move into office at the end of 1988, his transition team received the report. Gray set up a team in the White House to develop the Project 88 proposal, working closely with EDF staff.

Their revisions to the *Clean Air Act* were kept secret from EDF's allies in the National Clean Air Coalition, who found out only after the bill, with its emissions-trading provisions, was announced. They were skeptical of an approach that focused on the needs of industry: Big Business was no longer the enemy; it was the solution. Worse, it was being given legal permission to pollute. Following the Coase-Crocker-Dales formulation, the legislation grants utilities a right to pollute at certain levels. If the utility emits less pollution than its assigned limit, the balance becomes a marketable asset or credit that can be banked for later use or sold to the highest bidder.[40]

By an overwhelming majority, Congress in 1990 amended the *Clean Air Act* to establish a market for electric utilities to trade rights to emit sulphur dioxide. The act came into effect in 1995. Within a very few years, the *Clean Air Act*'s emissions-trading market was credited with cutting sulphur dioxide emissions and saving money. It was so successful, its supporters claimed, that how could it not be the model for climate change negotiations, which began in Berlin in 1995. (Canada's proposal for a voluntary approach, the National Action Program on Climate Change, released in Berlin, is described in Chapter 4.)

The inconvenient truth, though, is that sulphur dioxide reductions were likely due to extraneous factors and not the market. During the same period the emissions-trading scheme was being installed, American railroads were deregulated and the cost of transporting low-sulphur coal over long distances, especially from the Powder River Basin in Wyoming and Montana, dropped dramatically. This is the source of the cheapest and lowest sulphur coal in the US. High-polluting plants designated for early entry into the market switched to Powder River Basin coal, cutting their emissions dramatically and saving money.[41]

Ronald Coase himself was dubious. Twenty-eight years after writing *The Problem of Social Cost* (1960), as the George H.W. Bush administration was set to foist emissions trading on the world, Coase pointed out that his doctrine couldn't apply to most real-world problems because of high transaction costs and inadequate information. His concerns were ignored.[42] Too much was at stake. Emissions trading captured the high ground and contrary information was ignored.

A comprehensive study of the acid rain program's first few years of operation published in 2000, *Markets for Clean Air*, presented an unapologetic defence of the program. It does at one point admit that the dramatic decrease in sulphur dioxide emissions and lowered costs were due to the fact that the most polluting plants (called the "big dirties") quickly switched to Powder River Basin coal.[43] Yet having raised that fact, the authors then ignore it and end the book with the statement that the program "clearly establishes that large-scale tradable-permit programs can work more or less as textbooks describe."[44]

This conclusion became the new reality — expensive technology is unnecessary because emission markets work! The inconsistencies between the book's conclusion and the impact of Powder River Basin coal may be explained by the fact that the book's lead author, Denny Ellerman, had been executive vice president of the National Coal Association during most of the time the acid rain program was being debated in Congress. A second author served on George H.W. Bush's Council of Economic Advisers during the development of the *Clean Air Act* amendments and was heavily involved in the design of the Acid Rain Program. And the third lead researcher was a member of the Environmental Protection Agency's Acid Rain Advisory Committee, at a time when the EPA was taking a lead role in promoting emissions markets.

The authors' preferences were clear. The book's epigraph is a quote from John Dales' *Pollution, Property and Prices*, pristine in its neoliberal intent: "If it is feasible to establish a market to implement a policy, no policy-maker can afford to do without one." Their hostility to public interest regulation is set out at the beginning. "Over the period since Dales produced his 1968 work, the alternative command-and-control approach to environmental policy, in which the design or performance of individual pollution sources is specified, has been applied to a wide variety of problems and has generally performed poorly, with excessive costs and, often, failure to achieve environmental objectives."[45] They present no references to support this claim of poor performance. They didn't have to.

By this time, the superiority of markets over public interest regulation had become self-evident. But the facts provided by the EPA present a different picture. In the twenty-five years between the 1970 act and the implementation of the new act in 1995, aggregate emissions of the six leading pollutants fell 40 per cent or 1.6 per cent a year. In the twenty years between the new act coming into force and 2015, emissions were down a further 31 per cent, or 1.6 per cent a year.[46] The market approach may have saved a bundle of money for polluting industry, but was no more effective than direct regulation. In fact, given the expanded use of Powder River Basin coal during the second period, the direct regulation of the first period had more impact than the market approach. But this fact was buried under the avalanche of good news about markets. And economists were the evangelists.

In the months leading up to the Kyoto Conference, 2,600 economists signed a widely distributed document, "The Economists' Statement on Climate Change." Signatories included eighteen winners and soon-to-be winners of the Swedish National Bank's Prize in Economic Sciences. Economists ranging from the American liberal left (such as Paul Krugman and Joseph Stiglitz) to the neoliberal right (Gordon Tullock, Ronald Coase student Oliver Williamson) signed a statement framing climate change as an economic problem with "many potential policies to reduce greenhouse-gas emissions for which the total benefits outweigh the total costs."[47]

The economists had good news for Americans. "For the United States in particular," the statement declared, "sound economic analysis shows that there are policy options that would slow climate change without harming American living standards, and these measures may in fact improve US productivity in the longer run." All that was needed was for the world to adopt market-based policies such as an international emissions trading agreement.

Coming eight months before the Conference of Parties 3, the effort was intended to support the Clinton administration's push for an international market in emissions permits. William Nordhaus, one of the first economists to address climate change in the 1970s, was an original drafter of this document. "Economists haven't been important players in environmental policy over the last thirty years," Nordhaus complained to the *New York Times*. "This time we could make a difference."[48] And they did. He and his economist colleagues helped Clinton get carbon markets into the Kyoto Protocol. More importantly, they made themselves indispensable in the climate change arena. Climate scientists, with their doom-and-gloom scenarios of devastating floods, heat waves and mass migrations could raise the alarm, but economists, with their soothing message of cost-benefit analysis and the magic of markets, would be the first responders.

CARBON TAX VERSUS CAP-AND-TRADE

As we've seen, to date there are two ways to price carbon — carbon tax (following Arthur Pigou) and carbon trading (following Ronald Coase, as interpreted by Thomas Crocker and John Dales). Both will increase the cost of polluting activity and both should lead to lower carbon emissions, or so the theory goes. (The question of whether either approach can reduce emissions enough to hold global warming to below two degrees is rarely raised.) But only one was chosen. When governments decided to act, they almost universally chose emissions trading or cap-and-trade, as it came to be called.

Cap-and-trade versus carbon tax is "a debate we never had," Australian consultant Barry Brook and academic Tim Kelly declared in their submission to an Australian Senate committee investigating carbon pricing.[49] Their observation applies world-wide. Emissions trading as the solution to climate change was the default position almost everywhere after its alleged success in dealing with sulphur dioxide in the US. In a chapter written by economists, the Intergovernmental Panel on Climate Change's second assessment report in 1995 claimed that "for a global treaty, a tradable quota system is the only potentially cost-effective arrangement where an agreed level of emissions is attained with certainty (subject to enforcement)."[50]

Despite the ultimate failure of this broad assertion, it became the economic profession's mantra. Markets were established in many jurisdictions. The European Union's Emissions Trading System, created in 2005, was "the shining pinnacle of this process," according to Jan-Peter Voss. The EU Commission describes emissions trading as "the key tool for reducing industrial greenhouse gas emissions cost-effectively." It is "by far the biggest such system in the world."[51]

During these years, the carbon tax was largely relegated to the margins.

This occurred without full consideration or debate, as Barry Brook and Tim Kelly point out. The pattern is curious for several reasons. Policy-makers, corporations and citizens are all very familiar with taxes. Well-defined structures for the imposition and collection of taxes have been in place for decades. Carbon markets, on the other hand, were new and had to be constructed from scratch. Neither business executives nor politicians, nor even economists, understood the complexity of carbon trading. Before most residents of Ontario had come to grips with the complexities of the province's emissions-trading system, it was cancelled in 2018 by the newly elected Progressive Conservative government of Doug Ford.

Despite these drawbacks, the consensus favoured cap-and-trade. Neoliberal ideology demanded market solutions over government-imposed taxes. Companies liked trading because it gave them more leeway in designing responses, with ample opportunity to profit from carbon markets. Companies especially favoured markets in which governments simply hand out licences to pollute rather than auction them off. Any potential revenue generated by the market will accrue to industry, not government or taxpayers. And politicians liked cap-and-trade because it wasn't as politically toxic as a new tax.[52]

Early efforts to adopt a carbon tax were vigorously rebuffed by industry. But carbon markets eventually came into disfavour, although many jurisdictions as of 2018 still rely on them. They rarely deliver real emissions reductions. Much of the carbon reduction noted in the European Union's Emissions Trading System resulted from companies switching from coal to gas, and had little to do with the market.[53] The markets can provide windfall profits for polluters, and who, besides the companies themselves, wants that? They are susceptible to fraud and corruption.[54] Finally, they fail to provide incentives for the private sector to actually invest in low-carbon technologies.[55]

Yet society will be stuck with these markets forever, with little transformation into a new way of powering its activities. Even the originators of the market approach, John Dales and Thomas Crocker, both cautioned against trading in carbon emissions. Markets can work where a few polluters create a discrete pollution problem, they argued. The American acid rain problem was limited to one country with a limited number of emitters. Carbon emissions, in contrast, are a global problem with myriad sources. "I'm skeptical that cap-and-trade is the most effective way to go about regulating carbon," Crocker told the *Wall Street Journal* in 2009, as the US Congress was considering sweeping cap-and-trade legislation (that never passed). He preferred an outright tax on emissions because it would be easier to implement and would allow flexibility for unanticipated problems.[56] And Dales said in a 2001 interview that cap-and-trade "isn't a cure-all for everything. There are lots of situations that don't apply."[57] In a 1968 article on emission trading, he had admitted that a carbon tax could achieve the same outcome as his proposed trading scheme, but it would be difficult to set and continually adjust the tax to achieve the desired outcome.[58] Dales's comment didn't have much impact at the time.

Economic consensus on a carbon tax began to coalesce in 2006 when American conservative economist Greg Mankiw launched the Pigou Club. Mankiw was coming off his term as head of George W. Bush's Council of Economic Advisers. His goal was to bring together leading economists, media commentators and political figures to promote a carbon tax. Mankiw opposed cap-and-trade because, he claimed, too often allowances to pollute are given away freely rather than being auctioned off — in effect, gifting polluters. But even cap-and-trade is better than "heavy-handed regulatory systems," he thought.[59]

Mankiw collected economists and others from a wide political spectrum, indicating the emerging agreement on carbon taxes: progressive Dean Baker, liberals Paul Krugman and Joseph Stiglitz, conservatives Kenneth Arrow, William Nordhaus and neoliberals Tyler Cowen and Gary Becker. Among the non-economists: Bernie Sanders, Al Gore, Republican Senator Lindsey Graham, billionaire Bill Gates and rocker Neil Young. Mankiw later joined up with a group of Republican "elder statesmen" to promote a carbon tax, including former secretaries of State James Baker III and George Shultz and former secretary of the Treasury Hank Paulson.[60] Support for the carbon tax had become widespread and non-partisan.

Yet even though the vast majority of economists favoured a carbon tax, there was little likelihood of such a tax coming out of the 2015 Paris talks. In contrast, while cap-and-trade was supported by a mere handful of economists, it was far more likely to emerge from the talks, which is what happened, thanks in large part to the efforts of BP and the International Emissions Trading Association, as Chapter 3 related.[61] So economists ceased debating carbon tax versus cap-and-trade and opened their arms to both, as the column from the Ecofiscal Commission that opened this chapter declares: "either economy-wide carbon tax or cap-and trade systems."

The debate then shifted to the question of what to do with revenues collected from the tax. Conservatives wanted the proceeds to be returned to taxpayers as cuts to income and corporate taxes. This is how the Gordon Campbell carbon tax was supposed to work in British Columbia. Tax the things you want less of — pollution — and cut taxes on things you want more of — income, profits. If carbon taxes go into government coffers, as Mankiw and his fellow conservatives seemed to fear, it will lead to ever-larger government. They were missing the full picture of Pigou's concept: Tax negative externalities and subsidize positive ones. The logic is that revenues collected from the tax should be used to finance technologies and actions that reduce emissions.

The case for revenue neutrality was at bottom an attack on government, a long-standing neoliberal strategy. As *The Globe and Mail* columnist Jeffrey Simpson wrote, "governments want to take the money from carbon pricing and then decide how to spend it . . ." Fundamentally they didn't trust the market to send signals to individuals and corporations to adjust behaviour. The government is going to do it for them, and at a higher cost.[62] A revenue-neutral carbon tax is neoliberal in intent. Here, the role of government is to set up the system, make sure it works properly, and then get out of the way. Neoliberals will vigorously attack any efforts to bring redistributive elements into the carbon tax system.

In 1989, just after climate change penetrated public consciousness, Greenpeace International declared to the energy committee of the British House of Commons that the ultimate objective of public policy should be to "[reduce] carbon emissions to the level where natural processes return carbon dioxide concentrations to natural levels."[63] Thanks to the preoccupation with carbon markets perpetuated by economists over thirty years, this aspiration can likely never be accomplished.

CHAPTER 7
FROM ENVIRONMENTAL CRISIS TO BUSINESS TRANSITION

In the years following the adoption of Kyoto in 1997, responses to global warming were mostly heading in the direction desired by business. Denial was still confusing publics about the veracity of climate change. Markets had been embedded in Kyoto and in official thinking. And voluntarism was still the default business response. But three events occurred in the crucial 2006 to 2008 period that threw the established order on its head: Nicholas Stern produced his report calling for serious spending on climate change abatement sooner rather than later; the Intergovernmental Panel on Climate Change released its fourth assessment report, asserting that "warming of the climate system is unequivocal" and "very likely due to the observed increase in anthropogenic greenhouse gas concentrations"; and the world was hit by the global financial meltdown.[1] These three events crowded in on one another and caused business to shift in a new direction.

THE OLD ORDER CARRIES ON
With carbon markets embedded in Kyoto, BP officials fanned out through the European Union and business committees to solidify support for trading.[2] Working closely with British government officials, BP hosted the UK Emissions Trading Group, a coalition of thirty oil and gas producers and

electric utilities, to develop a domestic trading scheme.[3] BP and the World Business Council for Sustainable Development then helped organize the International Emissions Trading Association, an organization of firms keen on emissions trading that became a central node for pro-trading forces, comprising fossil fuel companies, banks, law firms and consultancies that promote carbon markets. As of 2018, Canadian members include Enbridge, TransCanada Corporation, Suncor Energy, Capital Power and Ontario Power Generation. Other members were major operators in Canada, such as BP, Chevron, Royal Dutch Shell and Total. Before Kyoto, emissions trading was perceived as "a licence to pollute" for industry and as allowing industrialized countries to escape emission reductions. After Kyoto, emissions trading was simple common sense. It was a notable achievement for the pro-trading coalition.

Nor did business let up on advocating for the voluntary approach. One strategy was to sponsor parallel conferences in locations where UN conferences of parties (COP) were occurring. In Kyoto, in 1997, before the official conference began, the World Business Council for Sustainable Development co-sponsored (with the International Chamber of Commerce) a two-day conference titled "Voluntary business initiatives for mitigating climate change." The theme of this meeting was that efforts to tackle climate change should be voluntary, flexible and based on market principles. "Direct regulations are too strict and counterproductive," Michael Kohn, honorary president of the Chamber's Energy Commission and head of a Swiss power plant holding company, told the audience.[4]

The World Business Council for Sustainable Development and the International Chamber of Commerce continued to press for voluntary solutions, using their special access to government decision makers. In 2002 at the earth summit in Johannesburg, South Africa that marked ten years since Rio (Rio+Ten), the NGO Christian Aid made a blistering attack on the business community when binding environmental regulations on companies were once again dropped from the agenda in favour of voluntary codes. The draft plan called only for the "promotion of corporate accountability and responsibility and the exchange of best practices," the British-based international anti-poverty organization claimed. "Business has greater access and influence than any other group and we are concerned that the agenda is being unduly skewed towards the wish lists of companies and away from those of the poor," the agency wrote.[5]

If the fossil fuel industry wasn't giving up on the voluntary approach, it wasn't done with denialism either. While BP and its allies were pushing ahead with markets and the voluntary approach, ExxonMobil and its allies were continuing to back neoliberal think tanks and the handful of scientists they supported that were still denying that global warming was even happening (or, if it was, they held, it would be beneficial to humanity). Interfering in the production of carbon dioxide would destroy the economy, they alleged, because how could government regulators know better than the market how the economy should operate. The deniers even turned their guns on market-based climate policy of the kind promoted by BP. In their view, emissions markets constructed by government bureaucrats are "unnatural markets," because the state imposes an artificial scarcity of goods (rights to pollute) upon participants, restricting the natural self-regulating capacity of the market and thus violating neoliberal principles.[6] It's just more state intervention.

THE *STERN REVIEW*

The attack on the market principles embedded in Kyoto fed into George W. Bush's hands. The US under Bill Clinton had signed Kyoto but Congress wouldn't ratify it. Then Bush withdrew the American signature. Denial and the voluntary approach retained their currency under his administration, which was the target of most denial messaging. That allowed Bush to claim, with the backing of ExxonMobil, that tackling climate change would ruin the economy and kill millions of jobs, without any benefit to the environment. So in July 2005, Bush formed a regional pact with Australia, India, China and South Korea. They were the Kyoto rebels: The US and Australia didn't sign the Kyoto Protocol; China and India were considered developing economies and not limited by the treaty; and South Korea was a major coal exporter. The Asia-Pacific Partnership on Clean Development and Climate was seen as an attempt to destabilize negotiations on a successor to Kyoto, which was set to expire in 2012. Like the Rio+Ten conference, it avoided greenhouse gas reduction targets and emphasized the exchange of cleaner technologies.[7] Bush kept the initiative secret from his supposed comrade-in-arms, UK Prime Minister Tony Blair until it was officially announced.

The British were not amused by this "snub." In response, Gordon Brown, then the UK's chancellor of the exchequer, launched a far-sweeping investigation into the damage global warming would do to the UK and the world. Take that, Dubya!

Brown asked Nicholas Stern, a former chief economist at the World Bank and then second permanent secretary at Her Majesty's Treasury, to head the investigation.[8] Stern and the team he led released their report in October 2006, and it attracted enormous attention, launching a new era of thinking about global warming. It was a direct rebuttal to Bush and his anti-Kyoto allies (and to Canadian prime minister Stephen Harper, who would eventually withdraw from Kyoto). In a powerful statement of the costs and benefits of taking action, the report, published as the "*Stern Review* on the Economics of Climate Change," predicted that if society does nothing — as George Bush and Stephen Harper seemed to want — "the overall costs and risks of climate change will be equivalent to losing at least 5 per cent of global GDP or more if a wider range of risks and impacts are taken into account." If, however, society takes prompt, strong action and moves to a lower carbon future, the costs can be limited to around 1 per cent of global GDP a year.[9]

The *Stern Review* was barely off the presses when the neoliberal denialist machine responded in predictable fashion. The Competitive Enterprise Institute's Myron Ebell claimed the report's estimates for reducing greenhouse gas emissions "are laughably rosy, while the assumptions about the impact of global warming are ridiculously overblown," he declared.[10] Many neoliberal think tanks amassed quotes from denialists and wrote articles slamming the "radical climate alarmism" lurking in the report.

Denialists continued to speak to their still sizable base, but the *Stern Review* was seen in business and political circles as a harbinger of things to come. As well as the standard cost-benefit analysis that attempts to assess the impact of climate change on the economy, Stern popularized the emerging perspective that the transition to a low-carbon economy could open new opportunities in all business sectors.[11]

It was the profitable solution business had been seeking for more than a decade. By 2050, Stern predicted, markets for low-carbon energy products could be worth US$500 billion a year, while investments in new power generation could exceed US$13 trillion and the market for emissions reductions could surpass US$2 trillion a year. "Tackling climate change is the pro-growth strategy for the longer term," he wrote, "and it can be done in a way that does not capture the aspirations for growth of rich or poor countries."[12]

This was a way forward that didn't depend on what you thought about climate change. As Jeremy Warner, the long-serving business editor of the London-based *Independent* newspaper observed after meeting with numerous business leaders, "more and more [they] are coming to recognize that doing something about global warming is key to having a viable business future," he wrote the week the *Review* was released.[13] For some businesses, if they wouldn't change voluntarily, the changes would be forced on them by government regulation. Other businesses would be attracted by the opportunities provided by green investment, Warner suggested. He could point to Stern's numbers for support.

FROM ENVIRONMENTAL DISASTER TO BUSINESS RISK

The Confederation of British Industry, Britain's most powerful business lobbying and public relations organization, summed up business thinking at the time in its publication "Climate change: Everyone's business." In the foreword, Ben Verwaayen, chair of the confederation's climate change task force and chief executive of the BT Group, the former British Telecom, asked:

> Are we sure that climate change exists? I am sorry, but that is not a question for us. The best question for the business community is whether we can be certain that climate change presents a substantial risk: a risk that will have a profound impact on society and the economy? To this the answer is clearly "yes." And so, as with all substantial risks, it is vital to mitigate the danger.[14]

Science historians Vladimir Jankovic and Andrew Bowman point out that Verwaayen is signalling a shift away from worrying about the environmental disaster resulting from climate change to worrying about the business risk.[15] Climatologist James Hansen called climate change a "planetary emergency"[16]; Harvard Business School professors Andrew Hoffman and John Woody called it a "market transition."[17]

Hoffman and Woody's framing was winning the day. It was an early indication that business was gaining the upper hand in the battle to define what kind of a problem global warming really was. If global warming were a planetary emergency, as Hansen claimed, then a massive government-led response would be the correct one. Think of the mobilization of resources by governments in the US, Canada, the UK and other allies to fight the Second World War. Then expand that to encompass the entire world. Global business would be shaken to its core. If, however, global warming was simply a business risk, albeit one of considerable magnitude, then business could take care of it, with a little help from its friends in government.

The 2007–2008 financial meltdown provided business with a golden opportunity to put the new framing into practice by redefining global warming as a market transition. Instead of reining in capitalism's rampant excesses that led to the most severe economic collapse since the Great Depression, governments provided hundreds of billions of dollars to bail out financial institutions and auto makers and re-established the system almost as before.

In contrast, government responses to the global warming crisis were muted and confused. Let markets deal with this, governments seemed to have concluded. For its part, business came up with a new strategy for capital accumulation — the green economy. It would lead us out of the crisis. Recovery could occur by expanding the green economy, which meant investing in decarbonization of production and consumption within the existing market system. But market-based strategies would simply reinforce existing property rights and power relations, and exacerbate other environmental problems that were caused by market activities in the first place. They were likely to worsen inequality and may not even yield net global environmental benefits.

Investment bank Lehman Brothers was one of the first global financial institutions to study the potential impacts of global warming on major business sectors: There were many challenges but also many business opportunities, as firms adapt, or fail to adapt, to the new environment, the study concluded.[18] Ironically, Lehman had more impact on the shift to the low-carbon economy than it could have imagined. It wasn't too long before the firm filed for bankruptcy protection, accelerating the global financial meltdown.

Business and political elites turned to the new framing — climate change as market opportunity — as the way out of the crisis. And it wasn't just business. At the third Investor Summit on Climate Risk held at UN head-quarters in New York in February 2008, UN Secretary-General Ban Ki-moon addressed 500 investors, Wall Street bankers and corporate chief executives:

> *You are here today because you recognize climate change as*
> *an opportunity, as well as a threat. You understand that the*
> *shift to a low-carbon economy opens new revenue streams and*
> *creates new markets. You see the chance to usher in a new*
> *age of green economics and truly sustainable development.*
> *And you seek to address climate change in ways that are both*
> *affordable and promote prosperity.[19]*

With banks crashing around the world and news about global warming becoming more alarming, it didn't take long for economic and climate crises to be joined together. In 2008 and early 2009, private sector and academic reports called for stimulus spending on a green economy as the way to solve both problems. One radical proposal came from the UK-based Green New Deal group, with members from the New Economics Foundation, Friends of the Earth, the British Green Party and the *Guardian* newspaper. Their inspiration, they explained, derived from American president Franklin D. Roosevelt and his New Deal, which he launched after taking office in 1933 in the midst of the Great Depression.

The overarching principle in the New Deal was a government-regulated economy in which the power of financial capital was tethered and the economy boosted through massive spending on infrastructure and social programs. The Green New Deal proposed similar massive investment in renewable energy, which would create thousands of jobs, and the forcible "demerging" of large powerful banking groups, with tighter regulation of the financial sector:

First, it outlines a structural transformation of the regulation
of national and international financial systems, and
major changes to taxation systems. And, second, it calls
for a sustained programme to invest in and deploy energy
conservation and renewable energies, coupled with effective
demand management.[20]

Little of this happened. Banks were not broken up, the financial sector was not reregulated, nor did massive investment go into renewable energy. Instead, investment poured into the Alberta oil sands and other large fossil fuel resources such as Brazilian offshore oil, and carbon exchanges opened for business. Nonetheless there was still an undeniable move — more of a trickle than a flood — in the direction of what came to be called the green economy.

Several months after the Green New Deal document was published, the UN Environment Programme launched its own global green new deal. The international body was concerned that when investment picked up again it would go into the traditional extractive economy. Instead, the UN Environment Programme wanted investment to seek out transformative activities such as clean technologies and natural infrastructure such as forests and soils.

Unlike the Green New Deal group, this report wasn't written in the language of progressive economic change à la FDR, even though it invoked the American president. Instead it was framed in a business-friendly dialect. The role of government is to "send the right market signals to investors, entrepreneurs and consumers, so 'we move from mining the planet to managing and reinvesting in it,'" said Achim Steiner, the UN Environment Programme's executive director (a position occupied by Maurice Strong thirty-five years earlier).[21] Edward Barbier, an American economist who helped write the report, explained that the report recommended spending 1 per cent of global GDP on green initiatives, but "money alone is not enough; spending must be accompanied by domestic and international policies — from removing perverse agricultural, fishing and energy subsidies that deplete environmental resources to taxing or trading carbon emissions."[22] This was no new deal, but the same old deal business had promoted for a decade.

And the deal was sealed six months later in a report from HSBC, one of the world's largest banks, in a report titled "A climate for recovery." The report urged governments "to use low-carbon growth as a key lever for economic recovery." In its analysis of government recovery plans, the bank asked five questions: "is the green stimulus large enough, when will it materialise, is it really green, how many jobs will be created and how effective will it be in mobilising private investment?"[23] The bank didn't ask questions about the impact of the stimulus on cutting carbon dioxide emissions. As the Confederation of British Industry's Ben Verwaayen said, "that is not a question for us." Investment in the low-carbon economy had been delinked from reducing carbon dioxide emissions. Now business could pick up the ball and run in a direction of its own choosing.

By 2009, many major global corporations were supporting action, but of a limited nature. The neoliberal-inspired Copenhagen Consensus was signed by 500 companies. It stressed that "developed countries need to take on immediate and deep emission reduction commitments." But forget about government regulation, carbon taxes or ways to actually reduce emissions. The Communiqué identified the need for credible measurement, reporting and verification of emissions "as prerequisites for a robust global greenhouse gas emissions market." Carbon had been transformed into a commodity that could be bought and sold.[24]

COST-BENEFIT ANALYSIS

To assess the impact of low-carbon spending on the economy, economists offered cost-benefit analysis, which was the foundation of the *Stern Review*. Coming off their success in securing control of the acid rain problem, economists worked tirelessly to gain entry into climate change policy-making. A year before Kyoto, Swedish National Bank's Prize in Economic Sciences-winning economist Kenneth Arrow, and ten prominent colleagues published a short paper in *Science* magazine arguing that cost-benefit analysis could provide "an exceptionally useful framework" in improving environmental policy outcomes.[25] Cost-benefit analysis "[sets] out the social costs and benefits of an investment project and [evaluates] whether or not the project should be undertaken."[26]

Cost-benefit analysis seems simple enough. The costs of controlling

119

greenhouse gas emissions are weighed against the benefits of avoiding climate-related damages to human welfare. Given a careful analysis, one should then be able to calculate "optimal" long-term policy choices. How much should we spend today to achieve a certain outcome tomorrow?[27]

Cost-benefit analysis had been used since the 1920s to help make decisions about dams, railroads and flood-control projects. In the 1950s, economists redefined cost-benefit analysis as a predominantly quantitative exercise and began applying the technique to a wide array of projects. The government of Canada used cost-benefit analysis to help evaluate government programs in the 1960s. Several economists applied it to greenhouse gas control during the 1970s. In 1979, Yale economist William Nordhaus produced the first full cost-benefit analysis of greenhouse gas reductions in relation to energy growth. He "discounted the greenhouse effect as not being significant enough to provide major constraints on energy growth, at least in the short term," one historical geographer reported.[28]

Economists began assessing climate control as if it were an appraisal of an investment project.[29] They used cost-benefit analysis at first to provide justifications for policy-makers to accept or reject specific policy proposals. Policy makers could choose option A because it cost less and delivered more benefits than option B, or so the cost-benefit analysis approach claimed. Later, economists were able to elevate cost-benefit analysis from being merely a tool to assess proposals, to providing the goals of climate change policy — the means to an end became the end itself. Never mind what society and scientists say, the cost-benefit analysis would tell us what to do.

In this process, economists took over the policy area, pushing scientists to the periphery. The Council of the European Union utilized cost-benefit analysis in its 2005 report on climate change. The report claimed that "the benefits of limiting the global temperature increase to 2 degrees Celsius outweigh the costs of the emissions cuts necessary to stay within this increase . . . A 2.5 degrees Celsius rise could cost as much as 1.5 to 2.0 per cent of global GDP in terms of future damage."[30] It sounds persuasive until one reaches the admission that "a lot of the damage cannot be readily expressed in monetary terms," such as loss of human life or biodiversity.

How much a human life is worth is an essential question in cost-benefit analysis. Is a human life in the North worth more than one in the South? In the 1995 Intergovernmental Panel on Climate Change report, economists based the dollar value of life on income differences, so that a rich person was worth fifteen times more than a poor one. William Nordhaus, in his cost-benefit analysis modelling exercises, speculated that a warming world will produce extra recreation benefits in the US and extra loss of life elsewhere.[31] That limitation didn't deter the EU. In cost-benefit analysis, there's no room for moral judgment. Once numbers have been attached to recreation benefits and life costs, the solution is clear: more golfing in the US and more dead people in the global south.

The biggest boost to a cost-benefit analysis approach came from the *Stern Review*. This report's main message is that cost-benefit analysis justifies prompt and strong action to reduce greenhouse gas emissions. Stern came to North America in February, 2007, to explain and defend his report, with stops in Toronto, Washington, D.C., where he testified before Congress, and New Haven, Connecticut, home of Yale University, where William Nordhaus held court. In a famous symposium, each laid out his version of cost-benefit analysis, although both supported a carbon tax. (And they were both members of the Pigou Club.)

Stern presented his case that the carbon dioxide already pumped into the atmosphere guarantees that the planet will keep getting warmer. The decisions people make in the next several decades will determine how much warmer. And the degree of warming will determine how many floods, wildfires, heat waves, crop failures, hurricanes and species extinctions will occur.[32] The most effective way of reducing emissions, Stern said, is an economic deterrent like a carbon tax set at a high level and commenced as soon as possible. "The later we leave it, the more difficult it will be, and the more risks you run."[33]

Nordhaus, with his leave-it-mostly-till later approach, was first up to challenge Stern. His main objection concerned the discount rate, the annual rate at which future costs and benefits are discounted.[34] Stern's rate is close to zero, meaning that we value the lives of future generations as much as we value our own. Therefore we need to cut our consumption — and our emissions — today to benefit those living tomorrow. Nordhaus, in sharp contrast, proposed a much steeper discount rate of about 3 per cent, meaning that benefits accruing in twenty-five years will be worth about half their current value. We should tax emissions less now and live with more warming than Stern would allow. Nordhaus based his critique on the assumption that the economy will continue to grow. Future generations will acquire more education and will be rich enough to redress any damage caused by climate change. Sea walls will protect cities, technology will solve problems of failing crops and spreading disease.

No one knows, however, if this good news scenario will unfold as Nordhaus predicts, because no one knows what life will be like on a planet that is significantly warmer. "If ever there was an example where there was uncertainty, this is it," commented Harvard economist Martin Weitzman, who was also at the debate. He was referring to the possibility, however slight, of disastrous temperatures and environmental catastrophes that could ruin the entire world. This possibility must dominate cost-benefit calculations. And this leads to the conclusion that we need strong climate policy. We just can't afford to follow the Nordhaus leave-it-till later approach.[35]

Economists lined up on either side of the debate — mostly with Nordhaus — and hurled polite and impolite missives at each other. But over the next few years, the two sides moved closer together as carbon emissions and global mean temperatures continued their relentless rise. Nordhaus had originally suggested a very low carbon tax of US$8 per short ton of carbon dioxide, which would take two decades to double and another three decades to double again.

By 2013, Nordhaus was recommending a starting tax nearly three times higher and with a much faster rate of increase — close to what Stern advocated.[36] The longer society takes to embark on serious action to bring emissions under control, the less will be the distinction between the two approaches. Nordhaus raised his estimate on the damage carbon pollution does to the economy again in 2017.[37] If he had been more open to the warnings sounded by climate scientists in the 1980s, economists might not have become such a drag on meaningful action over the following decades.

Yet how accurate are the models created by Stern and Nordhaus? A reality check, obviously, is not possible, since they were predicting the state of the world 100 years and more into the future. This drawback is well-recognized by economists, so they perform a sensitivity analysis of their models' underlying assumptions. These are technical, but try to answer a simple question: If you change the assumptions, does the model still produce the same output? For Stern's model, two sensitivity-analysis practitioners think not. Andrea Saltelli and Beatrice D'Hombres found that Stern's sensitivity analysis had "been used improperly and that — had it been used properly — it would have invalidated the analysis itself." Their arguments against the Stern analysis, they say, apply to Nordhaus as well. "Both authors pretend to describe the issue on terms of parameters and models which bear no tested relation to reality."[38] Even when used for discrete, well-defined projects, cost-benefit analysis can raise doubts about its simplifications and assumptions. But for global multi-century cost-benefit analysis such as climate change, the simplifications and assumptions are overwhelming.

Nonetheless, and despite the uncertainty of economic modelling, the *Stern Review* and the Nordhaus retort mobilized resource and environmental economists to grab a leading position in global warming debates. In 2007, the Association of Environmental and Resource Economists launched a new journal, the *Review of Environmental Economics and Policy*, intending to target an audience broader than just other economists.

CHAPTER 8
THE NEW GOSPEL

By 2007, *The Globe and Mail's* Jeffrey Simpson could ask if "the jig was finally up for the Canadian fossil fuel industry's long campaign of denial and delay against climate change?" He noted that for nearly twenty years, the industry had effectively countered all efforts to implement regulations and economic measures designed to reduce greenhouse-gas emissions. "Every time a Canadian government hinted or appeared ready to move in this direction, the industry successfully lobbied against action." Consequently, "there have been no carbon taxes, mandatory emissions requirements, obligatory targets on vehicle manufacturers, or cap-and-trade system." The industry's successful tactics have resulted in "voluntary measures and programs based on improving energy 'intensity,' or efficiency — something any sensible industry would try to achieve to reduce costs."[1]

It was a fair assessment. But as we saw in Chapter 7, everything quickly changed. The Intergovernmental Panel on Climate Change was preparing to release its Fourth Assessment Report, which would declare warming of the climate system to be a fact and "very likely due to man-made increases in greenhouse gas concentrations."[2] Most of North America was hit with an unseasonably warm winter. Al Gore had recently released his movie, *An Inconvenient Truth* that made global warming a popular discussion topic. And, perhaps most important for the business audience, Sir Nicolas Stern's report, the *Stern Review* on the Economics of Climate Change, was making the global rounds.

This report concluded that the benefits of strong early action on climate change far outweighed the costs of not acting.[3] When Stern came to Toronto in February 2007 with his call for carbon pricing, it was standing room only at an Economic Club luncheon at the Toronto Hilton. Clive Mather, chief executive of Shell Canada, introduced Stern to hundreds of Canada's business and political elite. "Life would have been much easier had we taken this up twenty or twenty-five years ago," he told the audience.[4] Stern could have directed these comments more specifically at his host, Clive Mather, whose company knew about the perils of global warming those same twenty or twenty-five years earlier and did nothing.

John Dillon, the Canadian Council of Chief Executives vice-president of regulatory affairs, was less than enthusiastic. "More and more businesses are recognizing that some form of cap, at some point, when the technology allows, is likely going to happen," he offered. A North American version of the European Union cap-and-trade system was a possibility, he admitted. "How quickly it's going to happen is another question."[5]

Dillon could afford to express doubt about Stern's call for immediate carbon pricing because a new discourse was taking hold, one that sidestepped the scientific evidence about climate change and therefore didn't require immediate, dramatic action. Harvard Business School professors Andrew Hoffman and John Woody wrote in 2008, "business executives . . . should not think of climate change as an environmental issue at all. Instead, you should think of it as a market transition."[6] Climate change was being reframed into an instrument of strategy that would lead to new areas of economic opportunity.[7]

The CEOs were soon working with this emerging discourse. Within a month of Stern's visit, the Canadian Council of Chief Executives head Tom d'Aquino assembled a task force of CEOs to consider how business should respond to climate change in this new reality. This group was led by d'Aquino, Alcan CEO Richard Evans and Suncor CEO Rick George. It included big oil heavy hitters like Murray Edwards of Canadian Natural Resources and executives from Petro-Canada, Canadian Oil Sands, Imperial Oil and TransAlta Corporation.[8]

Their report, titled "Clean Growth: Building a Canadian Environmental Superpower," came out in the fall of 2007. It presented a dramatic pivot for Canadian business. For a decade, the CEOs had expressed profound doubt about Canada's ability to achieve its Kyoto targets without ruining the economy. Now they were conveying optimism about a low-carbon future — a future based on a national energy plan, carbon reduction targets designed to protect corporate profits, investment in new technologies and appropriate rational carbon pricing.[9] "The climate change challenge offers immense opportunities for Canada," the document proposed.

The pivot followed almost exactly the lead taken by the World Business Council for Sustainable Development and European industry, demonstrating once again the co-ordination of industry policies at national and international levels. The clean-growth discourse in Canada commenced with this document, eventually becoming climate-change gospel. Business lined up behind it because it would create certainty for business planners. It also meant that growth could continue and even increase, although it would have to be clean, whatever that meant. Other sectors of society would soon line up behind the new discourse.

The report declared that market forces alone were unlikely to meet the challenge, so government intervention was necessary — not to regulate the industry's bad behaviour, but to raise energy prices as a means of influencing consumption, either as an emissions-trading scheme or environmental taxes. Investment in technology was key. Spurred by government incentives, it would help Canada become a leader in trimming emissions output. Above all, the policy had to feature a national plan. This was crucial. Calling for these particular instruments with which to fight climate change removed the threat of direct, vigorous government regulation from active consideration. In fact the word "regulation" appeared only once in the document.[10]

Tom d'Aquino became an evangelist for a national energy strategy, clean growth and carbon pricing. At a breakfast meeting of the Calgary Chamber of Commerce the following summer, he declared that "the debate should not be seen as a threat, but as a huge opportunity for all of us." Technology would be the tool that allowed companies, governments and consumers to meet targets around the reduction of greenhouse gas emissions. "Technology and investment in technology will get us to where we need to be. There has to be a solution . . . rather than curtailing growth."[11] And in a *National Post* op-ed, he wrote that the Canadian Council of Chief Executives had endorsed the use of price signals as a means of persuading businesses and consumers to reduce their emissions of greenhouse gases either through environmental taxes or a cap-and-trade system.[12]

A CARBON TAX CONVENIENTLY APPEARS

And they already had a carbon tax they could point to. It was in the same year 2008, the year the global economy tanked, that B.C. Liberal Premier Gordon Campbell brought in the nation's first carbon tax. Most economists loved the tax because of its design. It was widely applied, covering about 75 per cent of emissions in the province; it was applied at the point of purchase for fossil fuels burned for electricity, transportation and home heating. And most of them — but not all — loved the fact that it was revenue-neutral. This would become a controversial feature of the plan. In B.C.'s case, all revenues collected by the government were supposed to be met by equivalent cuts to other taxes. The tax started modestly at CAD$10 per tonne of carbon dioxide in 2008 and would rise in $5 increments to reach $30 per tonne in 2012. As the icing on the cake, low-income households, which spent a greater proportion of their income on energy, were assisted by a tax credit, while rural residents, who on average use more gasoline, received a direct grant.

There were some problems. The tax didn't apply to fuels exported from B.C., nor to fuel used by planes and ships travelling to or from the province, which were all major issues. Coal has a high-carbon content, but since coal mined in B.C. is mostly exported, it didn't attract the tax.

Nonetheless, there was good reason for the well-heeled in B.C. to celebrate. Thanks to Campbell's deep cuts to income and corporate tax rates when he took office in 2001 — 25 per cent to personal income tax and 20 per cent to the corporate tax and further cuts that accompanied the carbon tax — B.C. had one of the lowest personal income tax rates in Canada and one of the lowest corporate rates in North America.[13]

No wonder business liked the tax regime. The Vancouver Board of Trade called it a "smart carbon tax," and gave the budget that accompanied it an A grade.[14] *The Economist* was a big fan, too. "We have a winner," the conservative business magazine declared three years after the tax's introduction.[15] World Bank Group president Jim Yong Kim would call B.C.'s carbon price mechanism "one of the most powerful" examples of carbon pricing.[16] And Angel Gurria, secretary-general of the Organization for Economic Cooperation and Development, would characterize B.C.'s carbon tax as being "as near as we have to a textbook case" of an explicit carbon tax.[17] Canadian-based *Corporate Knights*, which calls itself "the magazine for clean capitalism," later awarded Campbell its CK Award of Distinction for his "courage and clarity" in unlocking "the power of markets to deliver outcomes that are better for the environment and the economy" although it didn't say better than what.[18]

Campbell had constructed a narrative of how he came to the brave decision to bring in the tax. It all started, he told *The Globe and Mail*, when he toured Beijing's Olympic stadium in 2006 (Vancouver was to host the 2010 Olympics) and was "met with a thick haze of pollution [that] left his eyes stinging and his throat raw."[19] It was a "visible manifestation of man's impact on the environment around him," he explained.

Then Campbell conferred with California governor Arnold Schwarzenegger, who warned him that climate change "threatens every person in the world." Campbell said he had initially been gripped by Al Gore's 1992 bestseller, *Earth in the Balance*, and read Australian climate scientist Tim Flannery's *The Weather Makers*. He perused the Intergovernmental Panel on Climate Change's 2007 report and quietly grilled scientists and economists such as Simon Fraser University energy economist Mark Jaccard.

A year after his Beijing visit, Campbell came up with the plan to reduce provincial greenhouse gas emissions 33 per cent below 2007 levels by 2020 (or 10 per cent below 1990 levels). In November 2007, the provincial legislature passed the *Greenhouse Gas Reductions Target Act* to enact the cuts.

Campbell's narrative of how he came to the decision to bring in a carbon tax was a self-aggrandizing account of a heroic political leader taking an enormous risk for the betterment of society and the environment. But there's another version of this story in which Campbell is not as green or as heroic.

Campbell nearly lost the 2005 provincial election because of lingering resentment in many parts of the province over his actions after his crushing

2001 victory: his surprisingly large tax cuts; his deep cuts to government spending; his attacks on public sector unions; and his privatization of BC Rail and other provincial Crown corporations. He may have seen the carbon tax as a way to co-opt the "green-minded swing vote," *The Tyee*'s David Beers suggests, as Schwarzenegger successfully did in California.[20]

The tax "shredded the delicate labour-environmentalist coalition" that B.C. New Democrats had worked so hard to craft. The New Democrats were slamming Campbell's "gas tax," enraging many environmentalists. The David Suzuki Foundation, Pembina Institute and ForestEthics called on the NDP to back off.[21] "People who had been involved in environmental causes that probably had no intention of voting for our government before," Campbell admitted to *Corporate Knights*, "realized that if they didn't support this kind of important long-term policy, then it would never be done again. And so they did support it," and gave Campbell his victory.[22] Suzuki called Campbell's win a "watershed moment;" Pembina, "an encouraging result."[23]

Environmentalists were still praising Campbell when he went to Copenhagen later in the year for the ill-fated COP15. At a gala awards ceremony organized by ten Canadian environmental groups to recognize "acts of climate leadership," Campbell received an Economy Wide Carbon Pricing award presented by environmental campaigner Tzeporah Berman, then of PowerUp Canada. Most of Big Green was present: Suzuki Foundation, Ecology Action Centre, Environmental Defence, Equiterre, ForestEthics, Pembina and World Wildlife Fund Canada.[24]

But the environmental groups changed their tune after Campbell retired and Harper appointed him as Canada's high commissioner in London. Campbell began lobbying European governments to accept Alberta bitumen as a clean source of energy.[25] The Harper government set up the pan-European oil sands advocacy strategy to operate from Canadian embassies across Europe. London, where Campbell was based, was the "team leader." The strategy included lobbying European decision-makers to weaken or undermine clean-fuel policies that would require Alberta oil sands products to be labelled as having higher emissions. The strategy noted that while Europe was not an important market for oil sands-derived products, European regulators had the potential to impact the industry globally.[26]

Berman was shocked after being informed of Campbell's oil sands lobby-ing. "I have never been so embarrassed," she told *The Tyee*. Environmentalists should have looked more carefully. While Campbell was touting cuts in car-bon dioxide emissions, he was building a 40-kilometre, billion-dollar free-way to open up the Lower Mainland as a gateway to Asia, paving over prime farmland and generating more pollution and carbon dioxide emissions. He was spending CAD$800 million to increase vehicle capacity on the Sea-to-Sky Highway between Vancouver and Whistler for the 2010 Winter Olympics. The expansion turned the resource town of Squamish, now just a forty-minute car ride from Vancouver, into a bedroom community with a vast scale-up in vehicle trips.

He was pushing hard to expand northern B.C.'s natural gas industry. He was supporting Enbridge's Northern Gateway pipeline, which was projected to generate fifteen million tonnes of carbon dioxide equivalent a year when the heavy oil was produced and another sixty million when the fuel was burned. And earlier, in 2002, with the premiers of Alberta and Ontario, he mounted a vigorous campaign to oppose Canadian ratification of Kyoto. None of these actions correlated with Campbell's vow to cut provincial greenhouse gas emissions by a third by 2020. Campbell's actions suggested his carbon tax was not driven by a desire to meet Kyoto's requirements.

That leads to a third version of the story, putting Campbell's tax into an international context. The background document for the 2007 enabling legislation was written by one Bruce Sampson, BC Hydro's senior vice-president of strategic planning and sustainability, who was seconded by Campbell to work on the plan. Sampson was a long-time provincial govern-ment senior bureaucrat before moving to Hydro.[27] One of his Hydro duties was to be the utility's representative on the World Business Council for Sustainable Development, which, as related earlier, led the world business lobby at the United Nations Framework Convention on Climate Change. As the Campbell tax was being developed, the World Business Council for Sustainable Development was pivoting from a voluntary approach-plus-subsidies to pricing. British Columbia would be a guinea pig for a carbon tax. No wonder its progress attracted so much attention. They were already watching.

THE STEPHEN HARPER DETOUR

The chief executives seemed to be doing it again, pushing public policy in the direction they desired, at least in British Columbia. This did not happen right away in the rest of the country. Stephen Harper, elected in January 2006, and prime minister for the next nine years, was a rabid cheerleader for Alberta oil sands projects and the pipelines needed to carry diluted bitumen to markets. Ironically though, his enthusiasm presented an insurmountable obstacle to Big Oil's ambitions. His plan for economic growth was to turn Canada into an "energy superpower." Nothing could be allowed to stand in the way of this lofty mission. All of this was to the good. But Harper's hostility to carbon taxes and to federal-provincial negotiations — key elements in the CEO agenda — prevented him from moving their plan forward. As always, he did it his way.

"Kyoto is essentially a socialist scheme to suck money out of wealth-producing nations," he famously wrote to Canadian Alliance members requesting funds to stop the Chrétien government from ratifying the "job-killing, economy destroying Kyoto Accord" in 2002.[28] The Canadian Alliance itself had deep fossil fuel industry roots. Created as the Reform Party under Preston Manning, the party was bankrolled largely by Alberta oil executives who held their organizing meetings at the Calgary Ranchmen's Club. As a frequent oil industry consultant, Manning was able to win their support.[29] Harper had been with Reform from the beginning and drafted the party's platform and statement of principles, which formed the party's policy bible, known as the "Blue Book."[30]

Oil industry influence continued to exert itself as Harper moved closer to the prime minister's office. His 2006 election war room had numerous ties to fossil fuel and other carbon dioxide-emitting industries. Long-time ally Ken Boessenkool, for instance, resigned his lobbying job at PR giant Hill+Knowlton to join the Harper election team, giving up ConocoPhillips as a client. Other lobbyists helping the campaign, such as communications aides Yaroslav Baran and Sandra Buckler, also ceased representing their clients. But during the campaign and after, their colleagues still represented Big Oil interests.

Harper came into office in 2006 attacking the Chrétien and Martin governments for ratifying Kyoto targets without a plan to reach them. Under the Liberals, emissions rose by 24 per cent. When added to the 6 per cent cut below 1990 levels Chrétien had committed to, the country was 30 per cent above target. Everyone agreed Kyoto was unachievable, Harper concluded, so it could be ignored. Instead he resurrected the phantom "made-in-Canada" solution while cancelling billions of dollars in federal spending to address climate change and promote energy efficiency.

THE MECHANIC ARRIVES

To execute his plan, Harper assigned Bruce Carson, one of his top aides, to be chief of staff to Environment Minister Rona Ambrose, a climate change denier, as new legislation was prepared.

Carson, who was known as "the mechanic" because he could fix anything, was brought in to ride herd on the effort and prevent the ultimate setback, losing a confidence vote.

Bruce Carson was a long-time party apparatchik with a "complicated past." The complication stemmed from the fact that in 1982 he was disbarred and served time in jail after pleading guilty to two counts of defrauding law clients. After serving his sentence, it was redemption time, at least until his next stint in jail. As well as a law degree from the University of Ottawa, Carson held a graduate degree in constitutional law from the University of Toronto. His first political job was doing research for various Tory senators, including former Senate Opposition Leader John Lynch-Stanton. He also served in the Progressive Conservative research office at the Ontario legislature.

His next stop was as director of policy research for then Leader of the Opposition, Stephen Harper. Carson came into Harper's office from the Progressive Conservative side of the merged party, having worked closely with PC party leader Peter MacKay to put together an interim policy document for the party. In Harper's office, he was appointed a senior policy adviser responsible for compiling the platform for the next election, which could come suddenly, given the fact of a minority Paul Martin government.[31]

The election did come and Harper eked out a minority victory. Carson went right to work as "legislative assistant" in the Prime Minister's Office. Such a description minimizes the key role he played on the priority files of energy, environment and Aboriginal affairs. Carson was soon at Environment Canada, working with a handful of officials to invent the entire air quality plan for Canada to the year 2050.[32] Ambrose was left high and dry as the Prime Minister's Office took charge of efforts to rebrand the party's environmental credentials. *The Clean Air Act* received a rocky reception in Parliament and Ambrose was shuffled out of her job shortly after.

Carson, meanwhile, continued his upward ascent. During his first year in the Prime Minister's Office, he was approached by the president and a senior climate scientist from the University of Calgary for funding for a school that would engage in clean energy research. The request was well received by Carson and the Prime Minister's Office, but perhaps not for the stated purpose. Harper's 2007 budget included CAD$15 million for this project, named the Canada School of Energy and Environment. On paper, it would be one of Canada's centres of research excellence and commercialization. In practice, it would be something quite different. A year later Carson was appointed executive director of the school after an international recruitment search, or so it was claimed. He had no training or experience in energy or environment. But he did know how to work Ottawa.

It was an unusual kind of appointment. Carson would be splitting his time between Calgary and Ottawa until after the next election. To the Liberals, it was a clear conflict of interest. "How can the prime minister's senior adviser possibly accept a position that's dependent on funding from the federal government, while continuing to work part-time in the Prime Minister's Office?" one Liberal MP asked.[33] Perhaps that was the intention, to create an arm's-length relationship with an organization that was doing the Prime Minister's Office's dirty work.

Evidence for that possibility could be seen in Carson's statement that one of his jobs at Canada School of Energy and Environment was to counter the dirty oil charge that environmentalists were hurling at the oil sands. Perceptive environmental commentator Andrew Nikiforuk points to the tight ties between the school and the Conservatives. Carson's deputy director, Zoe Addington, had served as policy adviser to two Tory cabinet ministers, Tony Clement and Jim Prentice. A former head of the Alberta Progressive Conservative Party chaired the school's board of directors.[34] Such contracting out of the Prime Minister's Office functions became standard operating procedure for the Harper government.

Carson was barely on the job before he changed the school's mandate to include working with government on energy, environment and climate change policy. Its goals became focused on protecting the industry from the dirty oil accusation. "We have to reframe the debate and get the message out that the oil patch is making changes in the way things are done. You are never going to satisfy the extremists, but there are a lot of other people who need to be told about what's really happening," he said.[35] The Carson/ Canada School of Energy and Environment gambit was just one Prime Minister's Office effort to resuscitate the oil sands reputation and demonize environmental opponents.

Disseminating messages about how clean the oil sands really are is not quite the same as undertaking clean energy research, but it's what Carson had shaped the Canada School of Energy and Environment to be. In this version of the Canada School of Energy and Environment, Carson would be working closely with Big Oil executives like Pierre Alvarez, outgoing Canadian Association of Petroleum Producers president, whom he had known and worked with for twenty-five years.[36] A month before he took on the job, he met with Petro-Canada CEO Ron Brennerman to discuss "energy." And the same month he met with Dave Collyer, still president of Shell Canada, but within the month would replace Alvarez as Canadian Association of Petroleum Producers president. Carson and Collyer would work closely together over the next three years on industry's plan to promote a national energy strategy through the industry-sponsored Energy Policy Institute of Canada — or at least until Carson ran into trouble in 2011.

THE HARPER DETOUR REACHES A DEAD END

John Baird was Harper's second environment minister — there were five over Harper's nine years in office. In April, 2007, Baird released the Harper government's second alternative to Kyoto, "Turning the Corner: An Action Plan to Reduce Greenhouse Gases and Air Pollution." Canada needs to do a U-turn, Baird said, "because we are going in the wrong direction" thanks to the inability of the Liberals to reduce greenhouse gas emissions. There would be tough industrial regulations that would require large emitters to reduce their emissions. But Baird was referring to emissions intensity — emissions per unit of energy output. Under this plan, total emissions could still go up, so this approach would never lead to overall reductions, which is what Kyoto required.[37] This corner was never turned, as Harper did not implement the plan.

Harper did bring in regulations banning new coal-fired power plants that did not capture and sequester the carbon they produced. But this requirement didn't apply to the many existing power plants that would be allowed to continue operating for their normal useful life of 50 years. No existing plants would be affected before 2030. Some provinces were pro-active on this file. Ontario phased out its coal-fired power plants by 2014, and Alberta's NDP government would bring in regulations to phase out coal-fired power by 2030. Harper did adopt the Barack Obama administration's new vehicle standards. But he did nothing to address the Alberta oil sands, the fastest-growing source of carbon dioxide emissions. And in December 2011, Harper pulled Canada out of Kyoto, claiming the country would have to pay billions of dollars for carbon offsets to meet its commitments, a result of the lack of action by Jean Chrétien, Paul Martin (and himself).[38]

Harper lent his support to his fossil fuel backers by demonizing opponents of energy projects, labelling them "radical environmentalists." He audited environmental charities; cut his government's spending on the environment; and muzzled his government's scientists. In the end, though, these moves were not what the chief executives wanted — a national energy strategy based on federal-provincial agreements, government subsidies for "clean" technology and a carbon tax.

Harper was ideologically incapable of delivering on these. He resisted agreements between the federal government and the provinces and he was dead set against a carbon tax. He shut down the National Round Table on the Environment and the Economy, a government advisory group set up by Brian Mulroney to provide independent arm's-length advice. He didn't like the advice it was providing, that government should adopt a carbon tax.[39] "Why should taxpayers have to pay for more than ten reports promoting a carbon tax, something which the people of Canada have repeatedly rejected," former environment minister Baird asked in Parliament as he defended his government's killing of the agency.[40] The program designed by the Canadian Association of Petroleum Producers and the Canadian Council of Chief Executives waited for nearly a decade until a more sympathetic administration was ensconced in Ottawa.

CHAPTER 9
THE MECHANIC AND THE ARCHITECT

Stephen Harper's intransigence about anything to do with federal-provincial agreements and carbon taxes meant that Big Oil had to wait out his time in office before it could implement its national energy strategy. But in the meantime it had work to do to bring together all corners of the oil and gas industry and other sectors of society to support the plan. The road from the Canadian Council of Chief Executives' 2007 "Clean Growth" declaration to the Trudeau government's 2016 Pan-Canadian Framework on Clean Growth and Climate Change twisted and turned through academia, federal and Alberta government bureaucracies, political parties, corporate lobbies, media and think tanks as Big Oil amassed support for the project. Unlike the hated National Energy Program of the 1980s, which was imposed unilaterally on the nation from Ottawa, this one would be created by Big Oil and its surrogates in provincial governments to further corporate interests.

A national energy strategy means different things to different people. For the progressive citizen group, the Council of Canadians, such a strategy should serve "two purposes: to ensure the stability of supply to Canadians and also to set out a plan to wean the country off its dependence on non-renewable resources."[1] These ideas were one-hundred-and-eighty degrees removed from Big Oil's ambitions. But Big Oil, and not the Council of Canadians, had the resources to capture the narrative.

A national energy strategy was first mentioned by industry executive Pat Daniel, CEO of Enbridge, in a speech to the Empire Club in Toronto in 2006. Daniel cast his comments broadly, intending to frame a national discussion. But many interpreted his remarks as applying to his company's Northern Gateway pipeline project, which proposed to transport bitumen from Alberta to the west coast. Northern Gateway was already running into opposition from First Nations along the proposed route.[2]

A national energy strategy could overcome these obstacles. But it couldn't be seen as emanating from industry. The idea had to shift from corporate advocacy to academic objectivity to be credible. That happened when Roger Gibbins, president of the Canada West Foundation, a Calgary-based think tank, published a report calling for such a strategy a year later and just a month after the CEOs published their clean growth declaration. Gibbins was a political scientist from the University of Calgary and an associate of the Calgary School of neoliberal and neoconservative political scientists, economists and historians who aided in the creation of the Reform Party and spread conservative ideas across the country in the 1990s and 2000s.[3] Alberta energy minister Ted Morton was a long-time associate of Gibbins in the political science department.

ROGER GIBBINS FORMULATES A NATIONAL ENERGY STRATEGY

Gibbins urged federal and provincial governments to develop policy for both energy development and climate change. They had to go together, he argued, because climate change can't be considered without discussing how energy is produced and supplied. And the debate needs to be based in the West, since that's where most energy resources are located, Gibbins insisted.[4] His proposal would yoke climate change to the energy policy cart and be steered by a western Canadian driver, ensuring the combined policy formation would head in a direction desired by Big Oil.[5]

Gibbins explained the project in the *Calgary Herald*. "If we get it right, Canada could position itself as a global leader in the pursuit of sustainable energy development and could enjoy untold economic prosperity — we could aspire to be, in Prime Minister Stephen Harper's words, 'a clean energy superpower.'"[6] Like so many corporate formulations of the task at hand that appeared during 2007 and 2008, climate change was being elbowed out of the limelight, while "clean" energy or "sustainable" energy became the story. As Gibbins later explained, in 2007 the talk was all about climate change but energy production should be top of mind; therefore we need a national energy strategy.

The government of Alberta realized it required "the protection, or cover, of a Canadian energy strategy." Pipelines were crucial to Alberta's ability to export its bitumen to American and Asian markets, he said. Getting them built would be much easier if federal and provincial governments were united behind the effort to do so.[7]

That the Canada West Foundation was the first to call for such a strategy is not surprising, given its location in the heart of oil country. The institute was formed by Calgary oil barons in the 1970s to promote western — i.e., oil — interests. Financial backing came from some of western Canada's wealthiest tycoons: Fred Mannix (Calgary billionaire in construction, coal and energy), James Richardson (Winnipeg stockbrokerage business, seed and fertilizer, grain elevators, oil and gas), Max Bell (newspapers, oil and gas) and Arthur Child (Burns Foods). Long-time Canada West chair Jim Gray was a co-founder and president of Canadian Hunter, a major fossil fuel producer, who led the fight against the National Energy Program in the 1980s. He was known as the "ayatollah" of conservatism, the person aspiring right-wing politicians had to see before launching their careers. Reform Party leader Preston Manning had been one such aspirant. Gray also backed Ralph Klein's 1992 Alberta Conservative leadership bid.[8] At Canada West, economic analysis is never far from politics.

Coming so close on the heels of the Canadian Council of Chief Executives' declaration, the timing of the Gibbins report may have been coincidental, but there were connections. A year earlier, Canada West and the CEO Council co-sponsored the North American Forum, a top-secret gathering of Canadian, American and Mexican military, political and business leaders at the Fairmont Banff Springs Hotel. The purpose was to focus on North American integration and security cooperation, and that included a North American energy strategy.[9] Peter Lougheed was the Canadian co-chair.

Gibbins and d'Aquino were there. (So were Suncor Energy's Rick George and TransCanada Corporation's Hal Kvisle. TransCanada was in the planning stage for its original Keystone pipeline, which would transport Alberta bitumen, perhaps from Suncor oil sands mines, to American destinations. Murray Edwards, whose CNR was gearing up to start its Horizon oil sands mine, was also in attendance.) At this stage the Canadian strategy may have been the North American strategy in disguise. Remember the effort four years earlier to promote a made-in-Canada solution to climate change? As discussed in Chapter 5, this was little more than a Canadian version of the George W. Bush plan for the US. A North American energy strategy, particularly in the context of security and integration, could simply mean the security of Canadian oil supply for American use. This would later change, as Canadian natural gas and oil sands producers went looking for new markets.

A second link between Canada West and the CEOs could be found in the funding of Gibbins's paper. It was the first publication in a Canada West project called *Getting it Right*, which was financed largely by the Max Bell Foundation, a Calgary-based charity established by Bell, a Canada West founder. Bell donated just under CAD$250,000 for the one-year project. Was it coincidence that Tom d'Aquino was on the Max Bell board of directors (2005–2010) when this grant was awarded?

The national energy strategy reappeared two years later in Winnipeg, when leaders from eleven think tanks agreed that a national dialogue on the role of energy in Canada's environmental future was very much needed. They called their agreement the Winnipeg Consensus. Sponsors of the meeting included Roger Gibbins of Canada West and the Business Council of Manitoba, whose CEO, Jim Carr, would later surface as Justin Trudeau's minister of Natural Resources, the official most responsible for implementing the strategy. The view of these eleven think tanks was that governments have been off-track in taking an "increasingly narrow regulatory approach to reduce greenhouse gas emissions."[10] Instead, governments should encourage investment in clean technology, infrastructure and a green economy. Get rid of regulation and let the market deal with the problem. And this wasn't even Big Oil sending Big Oil's message!

AN EPIC TALE

Six months after the Winnipeg meeting, a greatly expanded group met in Banff to develop a "truly Canadian clean energy strategy." What was said was not as important as who was there. The fossil fuel industry was out in force. Along with the think tanks from the Winnipeg Consensus were representatives from over a dozen fossil fuel companies, including Murray Edwards. Officials from the Canadian Energy Pipeline Association, Canadian Association of Petroleum Producers and Canadian Gas Association were also in attendance. Roger Gibbins was there, of course, but not Tom d'Aquino, who had been succeeded in the interim by John Manley, who had been industry and finance minister in the Chrétien government and was there. Several participants who would assume key positions in the Justin Trudeau government's clean growth initiative were present: Jim Carr; Gerald Butts, president of the World Wildlife Fund–Canada (one of a very few environmental groups in attendance), who would become Trudeau's principal secretary and played an important behind-the-scenes role in moving Trudeau's clean growth framework forward; and Marlo Raynolds, head of the Pembina Institute, who would be appointed chief of staff to Trudeau's Environment Minister, Catherine McKenna.

But the biggest news from Banff was the presence of six representatives of a new player on the scene, the Energy Policy Institute of Canada (EPIC). This organization was incorporated the same month the Winnipeg Consensus was reached, October 2009. It had the backing of Canada's largest fossil fuel companies like Shell Canada, Imperial Oil, Canadian Natural Resources, and Suncor Energy, pipeline companies TransCanada Corporation and Enbridge, plus the major fossil fuel industry associations and especially the Canadian Association of Petroleum Producers. Company CEOs paid CAD$50,000 each for a two-year membership. The Canada West Foundation and the CEO Council were both contributors to EPIC. EPIC was created to be industry's chosen vehicle to promote a business-friendly national energy strategy and have it adopted by federal and provincial governments. To do this exceptional political, corporate and media connections would be required. These were provided in spades by the organization's founders and directors.[11]

Front and centre was Bruce Carson as a co-chair of the organization. Carson was able to reach into the top echelons of the Harper government to promote the corporate plan. He arranged meetings with many senior officials, including Wayne Wouters, clerk of the Privy Council and the highest-ranking civil servant in Ottawa; Cassie Doyle, deputy minister in the all-important Natural Resources Canada department; and Nigel Wright, Harper's chief of staff.

Carson's interaction with Wright is revealing. Shortly after Wright assumed his post in January 2011, Carson emailed EPIC's draft energy report to him. "I don't think we have ever met — but we have a few mutual friends . . . thought I would share with you a report I just finished on energy . . . would love to meet with you at your convenience," Carson wrote. Wright replied that he'd "heard a lot of good things about you. Feel free to give me a call at any time. I'll read the report over the weekend."[12] In February, Carson briefed Wright about EPIC and reported to EPIC president Doug Black that Wright "seemed generally supportive . . . and now at least he has been briefed." Black responded to Carson in an email, "Excellent. Need Nigel on side." Black praised Carson's efforts. "We are making progress and you are the secret sauce."[13] As Gerry Protti, an EPIC founding director, explained, Carson was "probably seen as one of the most knowledgeable people in the country on how to navigate the federal process."[14]

Carson's work for the industry came to a crashing end in 2011 when the Aboriginal People's Television Network reported that Carson was using his contacts to secure water filtration contracts with First Nations for his twenty-two-year old girlfriend. He would be charged and convicted of illegal lobbying and influence peddling.[15]

If Carson was the mechanic, d'Aquino was the architect. He wasn't at the Banff Dialogue, but as the former head of the CEO Council, d'Aquino was the most important individual behind corporate efforts to gain control of the climate change agenda. He laid out the plan in his interview with RCMP investigators who were collecting evidence on Carson's role in illegally lobbying government officials. D'Aquino's agenda set industry's work plan. There were four elements: bring the entire corporate sector together; recruit sympathetic academics to produce favourable research; influence media opinion; and persuade government officials to embrace the strategy.[16]

D'Aquino had led the CEOs for thirty-three years. With the pivot to clean growth becoming embedded in informed policy thinking about climate change, and with EPIC positioned to lead the corporate effort, he turned the mantle of leadership over to John Manley, as mentioned previously.

Manley was perhaps the Canadian most qualified to lead the CEOs. This wasn't because Manley was an ex-Liberal cabinet minister and would be a good fit to face the Liberal government of Justin Trudeau. The CEOs appointed him in 2010, when Stephen Harper was prime minister. It was Manley's business advocacy that counted. He started as a corporate tax lawyer in Ottawa in the 1980s and became chair of the Ottawa Board of Trade, promoting business interests in the National Capital Region. He spent one term as a Liberal backbencher, and when Jean Chrétien took over in 1993, Manley was appointed minister of Industry, a position in which the occupant represents the interests of business in cabinet (just as the minister of Natural Resources represents mining and fossil fuel interests).

Manley did stints as minister of Foreign Affairs and Finance and served as deputy prime minister before running for the leadership of the party after Chrétien stepped down. He lost badly to Paul Martin and chose not to run in the 2004 election, in which Martin won a minority government.

Manley's first corporate assignment after his years in politics was as Canadian chair of the so-called Independent Task Force on North America, a project co-sponsored by the CEOs and the Washington-based Council on Foreign Relations. (Tom d'Aquino was the Canadian vice-chair.) The report recommended an economic union of Canada, the US and Mexico, something the CEOs had been promoting for years, but were never able to pull off. Manley was present at the secretive North American Forum where integration was the top agenda item. In short order, Manley joined big corporate law firm McCarthy Tétrault as senior counsel, was elected to the board of the Canadian Imperial Bank of Commerce (CIBC) (an obvious position for a former minister of Finance, who ultimately regulates banks) and was appointed a North American deputy chairman to the Trilateral Commission.

THE NATIONAL ENERGY STRATEGY PROGRESSES UNDER JOHN MANLEY

With Manley in the driver's seat, it wasn't long before corporate efforts bore political fruit. Care had been taken to ensure that this would not end

THE MECHANIC AND THE ARCHITECT

up as another Ottawa-imposed plan, the approach that doomed Pierre Trudeau's National Energy Program. The provinces owned the resources and had to be involved from the beginning. Ten months after EPIC set up shop, the Council of the Federation, the biannual meeting of provincial and territorial premiers, discussed the concept, but did not refer to it in the meeting's communiqué. A month later, the annual energy ministers meeting in Montreal did refer to the national energy strategy, which had been relabelled as a pan-Canadian framework to remove any suspicion this would be dominated by the federal government. "Ministers mandated their officials to identify areas of common interest as well as goals and objectives related to energy that will lead to greater pan-Canadian collaboration," the news release reported.[17]

The Harper government endorsed the strategy by July 2011, six months after Bruce Carson approached Nigel Wright, and two months after Harper won his long-sought majority in Parliament. "We have an interest in a pan-Canadian collaboration, certainly," Harper's new minister of National Resources, Joe Oliver, said in an interview with *The Globe and Mail*.[18] As noted above, though, Harper had little intention of following through on his government's endorsement.

That same month, the Canadian Council of Chief Executives set out a to-do list for energy ministers, who were to meet in Kananaskis, Alberta. The pan-Canadian framework must be driven by the private sector; all forms of energy — including coal and oil sands oil — will be needed to satisfy growing demand; the regulatory burden must be cut; public-private partnerships must be promoted; and carbon pricing must be part of the mix.[19] The energy ministers confirmed their support for this agenda. Just how far they had been sucked into the corporate vortex was revealed by a statement in their news release: "As global energy demand is expected to grow over the coming decades, Alberta's oil sands are a responsible and sustainable major supplier of energy to the world."[20]

Alberta's soon-to-be premier Alison Redford staked out the national energy strategy as her personal turf. In 2011, she was one of six candidates running to succeed premier Ed Stelmach. In response to a question from the Canadian Association of Petroleum Producers at an all-candidates forum, Redford said that Alberta must take a leading role in developing a national energy strategy or else the rules will be dictated to Alberta by others. "I don't want to be sitting back and waiting for the agenda to be developed," she replied. "If we articulate that and have all Albertans behind that issue, it gives us a much stronger voice at the table."[21]

With barely one month in office, she travelled to New York, Washington, D.C., Toronto and Ottawa to promote the Keystone XL pipeline and the energy strategy. In Ottawa, she had breakfast with Manley and former federal Environment Minister Jim Prentice, who was then a CIBC vice-president and would succeed her as premier.[22] Redford pushed the energy strategy onto the Council of the Federation agenda at the July 2012 meetings in Halifax. The CEO Council prepared a briefing document for the meetings. The premiers set up a working group led by Redford and the premiers of Manitoba and Newfoundland and Labrador to begin work on a strategy for "a sustainable energy future."[23]

EPIC released its strategy paper just a month after the premiers met in Halifax.[24] Carson was no longer with the organization, his illegal lobbying having caught up with him. EPIC's paper received little attention from the media, which was unfortunate because it was already influencing federal government policy. An analysis by the ForestEthics Advocacy Association discovered that through EPIC, the energy industry helped write the rules "that now restrict public participation on the environmental impacts of tar sands expansion projects."[25] ForestEthics documented the profound impact the EPIC report had in at least one crucial area of energy development — government regulation. EPIC recommended that the "federal government must develop regulations that restrict participation in federal environmental assessment reviews to those parties that are 'directly and adversely affected' by the proposal in question." It also recommended that, "the relevance and credibility of evidence presented for environmental assessments must be explained." This precise language could subsequently be found in the *Canadian Environmental Assessment Act 2012*, on the National Energy Board's website, and on the National Energy Board's Application to Participate Form, ForestEthics notes.[26] It's just one example of many in the report.

As the ForestEthics analysis demonstrates, the corporate strategy was well on the way toward political acceptance. The chief executives could ease up on the throttle and shift gears to other key issues — tax reform, enhanced free trade, jobs and skills training, innovation — as the initiative switched from the board room to the ground game. But the CEO Council's John Dillon would maintain a watching brief on the climate change file as chair of an informal energy industry coalition that "aims to positively influence government's climate change policy."[27]

UNIVERSITY OF CALGARY'S SCHOOL OF PUBLIC POLICY

The next item on d'Aquino's agenda was to move the strategy into the public realm, a daunting task given the low regard the public held for both politicians and corporate executives. If they were the only ones promoting the strategy, it would go nowhere. Something more was required, as *Alberta Oil* magazine discovered. In a national survey of Canadians it commissioned, the magazine reported that the oil and gas industry was viewed as the least trusted source of information on energy issues. Only 14 per cent of survey respondents believed energy company executives were a credible source of information about oil sands development, and just 10 per cent trusted energy company information about carbon emissions.[28] Federal and provincial governments fared little better as trusted and credible sources of information on these subjects. But there was a ray of hope. At 53 per cent for information on carbon dioxide emissions and 57 per cent for information on oil sands development, academics were considered more credible and trustworthy. So if industry executives couldn't speak for the industry, perhaps professors could.

Such thinking must have occurred to more than one oil industry executive. It was certainly what Tom d'Aquino had in mind when he laid out his master plan to recruit sympathetic academics to write favourable reports. It also occurred to energy lawyer James Palmer, who donated CAD$4 million to the University of Calgary to set up a policy institute — the School of Public Policy — that would produce research on energy and other topics and train students to work in policy, government and industry. With Palmer's money, the university recruited tax specialist Jack Mintz, CEO of the corporate-sponsored C.D. Howe Institute, to head the School of Public Policy. The new entity would not overlap with the work of the Canada School of Energy and Environment. After Carson took an enforced leave of absence in 2011, former Alberta Energy minister and former senior

Canadian Association of Petroleum Producers official Rick Hyndman was brought in to lead Carson's creation. It went back to its original pre-Carson mandate, to do the clean energy research that industry wanted.

As at C.D. Howe and the Canada School of Energy and Environment, corporate influence in the School of Public Policy has been heavy. (This is not to suggest that industry money could buy supportive academic research, but that academics sympathetic to business and conservative viewpoints may be recruited for such positions.)

The connections to one company in particular — Imperial Oil — are extensive. Mintz himself is an Imperial Oil director and a director of the Imperial Oil Foundation, which doles out $6-to-$7 million a year to organizations in communities where Imperial Oil operates, to build goodwill. Like all corporate directors, Mintz is obligated to advance the best interests of the company, as former Alberta Liberal leader Kevin Taft points out.[29] As head of the School of Public Policy, Mintz's loyalties seemed divided. Are the best interests of the corporation the same as the requirement that scholarly research be independent and objective?

James Palmer, who died in 2013, was one of Canada's top oil and gas lawyers, specializing in mergers and acquisitions.[30] He was senior lawyer in Burnet Duckworth and Palmer, Calgary's top law firm, where he mentored Murray Edwards in the early 1980s. Palmer was on the boards of numerous oil and gas companies, including Edwards's Canadian Natural Resources. For a few years Palmer lobbied the federal government for Imperial Oil and its parent company, ExxonMobil, promoting their oil pipeline proposals.

Imperial Oil's CEO, Tim Hearn, had just retired and joined the School of Public Policy's advisory council; his company subsequently donated $1 million to the school. Hearn's successor at Imperial Oil, Bruce Marsh, was a featured speaker at the school's kickoff conference. Jean-Sebastien Rioux, recruited to lead the school's Master's program, had previously headed Imperial Oil's lobbying and public relations efforts.

That the School of Public Policy would be top-heavy with fossil fuel industry support shouldn't be surprising, given its location in Calgary and given the makeup of the university's board of governors (although this would change after the New Democrats took office in 2015). Former board chair, Bonnie DuPont, was a senior executive at Enbridge, while other governors were from Suncor Energy and the Canadian Energy Pipeline Association. The university's president, Elizabeth Cannon, is a geomatics engineer, a director of Enbridge Income Fund Holdings, and a trustee of Enbridge Income Fund, which operates the company's oil and gas pipelines.

Cannon was paid CAD$130,500 in 2014 for her work as an independent director on top of the CAD$600,000 she earned from the university that year. She held the same duty to Enbridge that Mintz had to Imperial Oil. That issue was raised when Cannon intervened in university affairs to promote the Enbridge Centre for Corporate Sustainability in 2012.[31]

And then there's the EPIC connection. Preceding DuPont as chair was EPIC founder Doug Black. He had been on the university's board of governors since 2007, spanning the period during which both the School of Public Policy and the Canada School of Energy and Environment were established. As a senator elected by Albertans and appointed by Harper, Black was a one-person campaign to promote the national energy strategy in whatever forum he found himself.

Nor is it surprising the school has emphasized research supportive of the oil industry, given the industry's influence at least at the administrative level. The School of Public Policy was as concerned as *Alberta Oil* magazine about the lack of trust Canadians had "in the key voices that speak on energy issues."[32]

A study for the School of Public Policy by former journalist and government bureaucrat Dale Eisler claimed that, "with Canada being a major energy producer and exporter, the need for the public to be informed and engaged in energy issues is crucial."[33] Eisler made no mention of the even more crucial need for the public to be informed and engaged in the project to cut carbon dioxide emissions to near zero within the next few decades. It's the same old corporate framing about getting product to market in a "sustainable" fashion.

During its first few years, School of Public Policy research dealt frequently with fossil fuel issues. One paper written in collaboration with American oil pipeline consultants sounds the alarm that Alberta must get its bitumen to markets in the Pacific Rim as quickly as possible, or risk losing out to competitors. Canada would need to get on with the Northern Gateway and Trans Mountain expansion projects as quickly as possible, the paper urged, writing before Northern Gateway was shelved.[34] Another paper trashes the idea of green jobs, calling them "illusory" because only the market can create economically efficient jobs based on "the intensity of energy use and greenhouse gas emissions per unit of output," ignoring the nature or quality of the output.[35]

School of Public Policy research seemed to downplay negative impacts of energy development. One paper asserts that "Canada's plentiful resources are an indisputable blessing, and those critics of federal industrial policy who compare this country to illiberal and corrupt 'petro-states' are being either ignorant or deceitful."[36]

The Calgary School of Public Policy was doing its part to rehabilitate Big Oil's tarnished reputation. And after the Rachel Notley government brought in a carbon tax of CAD$30 a tonne in November 2015, a School of Public Policy author argued that revenues generated by the tax should be used to lower existing taxes and not be used to finance programs that could further reduce emissions and improve people's lives.[37]

UNIVERSITY OF OTTAWA'S SUSTAINABLE PROSPERITY NETWORK

While Imperial Oil was supporting academic research in the oil province, a parallel effort was under way in the nation's capital, with Shell Canada playing a lead role. It seemed as if both the federal and Alberta governments were being explicitly targeted. Academic validation of carbon pricing was a crucial building block in the national energy strategy. The Sustainable Prosperity Research and Policy Network, based at the University of Ottawa, appeared on the scene the same year the Canadian Council of Chief Executives published its clean-growth document and James Palmer donated CAD$4 million to the University of Calgary.

One purpose of the network was to promote market pricing as the best way to protect the environment. It recruited academics from across the country that held similar views. Sustainable Prosperity's first public pronouncement was to endorse Gordon Campbell's carbon tax as "smart public policy."[38] This opinion piece in *The Globe and Mail* was co-written by two leading academics in the network and the president of Shell Canada. Promoting carbon taxes and reframing the environment as a series of markets were core Sustainable Prosperity projects. Under "What We Do" — it has since been removed from its website — the organization claimed that:

THE MECHANIC AND THE ARCHITECT

The operating system of our modern world — the capitalist
market — is an incredible tool. It links billions of producers and
consumers every day, generating price signals that help people
around the world decide what to make and what to buy. But
when it comes to conserving Earth's natural environment, our
markets are badly broken: we don't pay the true environmental
costs of making, using, and getting rid of stuff.[39]

Sustainable Prosperity's mission has been to make markets work for the environment; bring the environment under the yoke of capitalism. An indication of this bias is the fact that Sustainable Prosperity has no environmentalists on its steering or research network committees or on its staff. The organization may have said its mandate is to make markets work for the environment, but there have been no voices speaking for environmental values within the organization. This task has been left to consultants, economists and business executives. The organization's steering committee illustrated the links among academia, government bureaucracy and fossil fuel industry.

As of 2018 Dan Gagnier of Winnipeg Consensus, Banff Dialogue and EPIC fame is a steering committee member. Gagnier joined the Trudeau team in 2012, the year EPIC's national energy policy was released. In 2015 he was the Liberal Party of Canada national campaign co-chair. Gagnier received his fifteen minutes of notoriety when he was fired from his position as Liberal co-chair four days before the election after it was revealed he was being paid to advise pipeline company TransCanada Corporation on the most effective methods to lobby a new Liberal government about the company's controversial Energy East pipeline, even while he was joining the Liberals on the campaign trail. Gagnier remained a consultant to TransCanada.[40]

Other steering committee members with fossil fuel connections included Mike Cleland, a former president and CEO of the Canadian Gas Association and a former deputy minister of Natural Resources Canada. (He was also at the Banff Dialogue.) Jack Mintz was an adviser while he was transitioning from the C.D. Howe Institute to the University of Calgary School of Public Policy.

Steering committee member Velma McColl had the closest connections to the fossil fuel industry and to the Liberal government.[41] She has been a long-time corporate lobbyist with over ten years at Earnscliffe Strategy Group, "the heaviest of heavy hitters on the Ottawa government relations scene," according to Glen McGregor of the *Ottawa Citizen*.[42]

McColl would show up at the Banff Dialogue. Before joining the lobbyist fraternity she spent seven years as an adviser to federal Liberal ministers of Fisheries, Environment, Health and Industry. Her list of fossil fuel clients is long. She lobbied for pipeline company Spectra Energy Transmission, General Motors Canada, oil sands operator Nexen, liquefied natural gas project developer BG Canada, Encana Corporation, Cenovus Energy, Nova Chemicals and Foothills Pipelines.

When the Liberals took over in late 2015, McColl added to her list of clients, picking up Shell Canada, a company she had represented briefly in 2006. One responsibility for Shell was to lobby the federal government on the development of future greenhouse gas emission regulations for the oil and gas sector. And she picked up B.C. liquefied natural gas industry clients. For Woodfibre LNG, her job was to "educate and inform" MPs and their staff about the Woodfibre project in Squamish, B.C. For the B.C. LNG Developers Alliance — indeed for all the liquefied natural gas hopefuls — she was to work with government to improve income tax regulations regarding accelerated capital cost allowances for liquefied natural gas, a government subsidy for the industry.

SUSTAINABLE PROSPERITY AND B.C.'S CARBON TAX

Promoting a revenue-neutral carbon tax has been a signature Sustainable Prosperity initiative. On the seventh anniversary of the inauguration of Gordon Campbell's carbon tax, Sustainable Prosperity economists wrote a glowing report on the tax's success.[43] Sustainable Prosperity liked the tax so much that in a blog post accompanying the report, it said that all Canadians needed to "celebrate" the tax's birthday and wish it many more. Why? Because it was "an economist's textbook prescription for the use of a carbon tax to reduce greenhouse gas emissions."[44] The Sustainable Prosperity researchers were in such a celebratory mood because through their research they were able to attribute fuel consumption and emissions reductions of between 5 and 15 per cent to the tax, and they could find no measurable impact on the overall economy, either positive or negative.

Studies undertaken by researchers without a connection to Sustainable Prosperity found quite different results. Consulting economist Marvin Shaffer questioned the assumptions underlying some of the studies cited in the Sustainable Prosperity paper that found up to a 15 per cent reduction in carbon dioxide emissions. "The higher end of the estimated impact of B.C.'s carbon tax," he wrote, "requires that people respond as much as four times more to price changes due to taxes as compared to price changes for other reasons, for example, an increase in costs," which is an unreasonable assumption.[45] Differences in trends may be impossible to attribute to the carbon tax alone, Shaffer writes. When the price of gas goes up because of shortages of supply, drivers continue with their same driving habits. But when the price goes up because of the carbon tax, drivers decide to drive less, Sustainable Prosperity's study claims. It was all a matter of salience, Sustainable Prosperity said. In salience theory, consumers pay more attention to some taxes than to others because they stand out. Consumers react more to the salient carbon tax than to the cost of gas. But the carbon tax is buried in the total cost of gas, while the price of gas is often reported in the media and can lead to line-ups at service stations. So salience cannot be a factor in this particular cost comparison.[46]

A 2014 study of the effect of the carbon tax on gasoline use in Metro Vancouver found that it reduced greenhouse gas emissions by less than 1 per cent and concluded that the tax "is relatively ineffective for directly reducing vehicular greenhouse gas emissions."[47] The Sustainable Prosperity paper doesn't cite this study, but all five papers it does reference as evidence of a positive link between the carbon tax and lower emissions have at least one author with a connection to the Sustainable Prosperity organization, indicating an organizational push to support the B.C. tax.[48]

Marc Lee, an economist with the progressive Canadian Centre for Policy Alternatives, has also disputed Sustainable Prosperity's numbers.[49] First, he showed that B.C.'s economy didn't perform as well as those of its neighbours, Alberta and Saskatchewan, between 2007 and 2014 (the full carbon tax period in his study) or from 2010 to 2014 (the period after the 2008 financial meltdown). Next, he demonstrated that B.C.'s total greenhouse gas emissions did fall between 2008 and 2010, but increased every year after that. During the first three years (2008–2010) greenhouse gas emissions fell 6.4 per cent, but this was due largely to the global financial meltdown and the ensuing economic collapse — emissions fell everywhere. Since 2010, though, emissions crept up 4.6 per cent by the end of 2015, or nearly 1 per cent a year. In total, greenhouse gas emissions in 2018 were down 2.1 per cent since 2007, a far cry from the target of 33 per cent below 2007 levels by 2020 mandated by Campbell.

Even when emissions per capita is used, Lee showed that emissions had started declining much earlier, in 2001 — well before Campbell's tax kicked in — and continued to fall until 2010, when they began to rise again. Lee concluded that the "impact of the carbon tax has been overstated by people who love carbon taxes." That claim would apply to Sustainable Prosperity, whose papers avoided the fact that the carbon tax's addition of 6.7 cents a litre to the price of gasoline paled in significance when placed beside the dramatic fall in crude oil prices in 2014 and 2015, which reduced gasoline prices by 30 to 40 cents a litre. The tax would have to be boosted to CAD$150 a tonne to restore the price to former levels. Then much higher taxes would need to be added to induce the reductions in fuel use required to meet Canada's greenhouse gas objectives.

But even if carbon taxes don't do much to lower emissions, they can serve another important purpose: financing measures that help offset the negative impacts of increased atmospheric carbon dioxide levels, following Arthur Pigou's recipe for neutralizing externalities through taxes and subsidies. To accomplish this, the B.C. tax would have to shed its vaunted revenue neutrality. Giving people back the money collected is not much of an incentive to reduce energy consumption, since people can spend their tax savings any way they like, such as taking a flight to Disneyland and adding carbon dioxide emissions. As Marvin Shaffer remarked, "if we are going to pay a tax in recognition of the costs we are imposing on future generations, those payments shouldn't be returned to us in the form of income or sales tax breaks." Instead, the revenues should be "directed to measures that benefit those who will be bearing the cost of what we do." Marc Lee provided one list: better public transit, energy-efficient buildings, zero-waste systems, renewable energy, forest conservation and stewardship measures.[50] Lee noted that it's more effective to use carbon tax revenues to invest in climate-action initiatives than it is to give the money back to individuals, because this will move society in a desired direction — a lower-carbon future.

But the views of Shaffer, Lee and others who support a carbon tax, but question the requirement that it be revenue neutral, were crowded out by the chorus of voices clamouring for revenue neutrality, a refrain that was largely sung by neoliberal sources and echoed by business and the commercial media as the only sensible option.

CHAPTER 10
PARIS: BIG OIL GETS ITS DEAL

Maurice Strong died three days before the 2015 UN Climate Change Conference got under way in Paris on November 30. It was the twenty-first Conference of Parties (COP), the annual session that commenced three years after Strong chaired the 1992 Rio Conference, which established the United Nations Framework Convention on Climate Change (UNFCCC). Led by Canadian Environment and Climate Change Minister Catherine McKenna and UNFCCC Executive Secretary Christiana Figueres, delegates paused in their deliberations on December 8 to pay tribute to Strong.[1] Former UN secretary-general Kofi Annan called Strong the "father of the world environmental movement."[2] And earlier, in May 2014, at an intimate eighty-fifth birthday party former Canadian governor-general Adrienne Clarkson and her husband John Ralston Saul hosted for their long-time friend, Clarkson claimed Strong "invented the environment."[3] They could point to substantial evidence to support their claims: the 1972 Stockholm Conference, the 1974 UN Environment Programme, the 1987 Brundtland Commission and the 1992 Rio Conference.

The *National Post's* Peter Foster had been a long-time critic of Strong and his multiple activities. Over two decades he'd written more than 100 columns that attacked the businessman-turned-diplomat. Foster did moderate his excoriation of the man after Strong's death, still calling him a "lifelong socialist," but also "the grand old man of the environmental movement." However, eighteen months later Foster was back at the name-calling: "The late great Canadian eco-meister Maurice Strong . . . may be gone, but his subspecies of power-hungry environmental alarmists is growing like an invasive weed."[4]

Those would be the 38,000 delegates at the Paris conference, which was billed as the make-or-break meeting for controlling climate change. COP21 may have been Maurice Strong's baby, but it was a very different creature than Foster imagined. Rather than being in the grip of power-hungry alarmists, Paris was guided to its inevitable conclusion by the veiled hand of Big Oil and its corporate and political allies. For nearly a decade, governments and the UN had been looking for a replacement for Kyoto, which was set to expire in 2012. The fix was supposed to happen in Copenhagen in 2009, but that conference collapsed in failure, a result of an inadequately prepared draft text, mistrust among developing countries and intractable opposition from the US and China.[5]

Paris, it was said, was the last chance. At the Durban COP17 in 2011, governments pledged to have a legally binding agreement by 2015 that applied to every country. This was it. So when Laurent Fabius, French Foreign Minister and President of COP21, brought down his gavel on December 12, signifying an agreement had been reached, "the assembled delegates erupted with cheers, a standing ovation, tears and warm embraces," American climate justice activist Nick Buxton reports. "Within moments the media wires and the non-governmental organization news releases were buzzing with reports of an 'historic' climate deal that marks an 'end to fossil fuels.'"[6]

Big environmental organizations were uniformly positive. Economist Nat Keohane of the Environmental Defense Fund, the organization most responsible for turning the world away from government regulation and toward markets, said that, "it means that we now have a chance — not a guarantee, but a chance — to put the world on a healthier path."[7] The deal provides a "strong framework for transparency and review. [This] is critical because it will provide the accountability needed to keep pressure on countries to meet their commitments. And it will send a strong signal to energy producers, world markets and governments that the future lies in clean energy." Environmentalist David Suzuki called the conference "a turning point . . . [that] may have been our last chance for a meaningful agreement to shift from fossil fuels to renewable energy before ongoing damage to the world's climate becomes irreversible and devastating."[8]

The goals of the agreement are to hold the increase in the global average temperature to "well below" two degrees Celsius above pre-industrial levels and to pursue efforts to limit the temperature increase to 1.5 degrees. Parties to the agreement aim to reach global peaking of greenhouse gas emissions "as soon as possible" and undertake rapid reductions thereafter in accordance with the best available science.

None of these elements of the agreement are mandatory, nor are the means to achieve them specified. The core concept is the nationally determined contribution that parties intend to make to reduce emissions. They should be "ambitious." But ambitious they are not because they will not achieve the target of two degrees, let alone 1.5.

Recognizing this problem, the agreement contains a pledge-and-review mechanism for governments to come together every five years to set more ambitious targets and to report to one another and the public on how well they are doing to implement their targets. These are among the few mandatory requirements in the treaty. But they are unlikely to work because there is no legal mechanism that requires governments to ratchet up their nationally determined contributions, nor is political pressure likely to work, as previous international agreements have demonstrated. (The World Trade Organization, for instance, uses a judicial system through a dispute settlement mechanism whose verdicts are backed by trade measures.[9]) As well, the agreement contains inbuilt delay — the first reporting period doesn't occur until five years after the agreement was adopted. The agreement authorizes the use of carbon trading to assist governments in achieving their nationally determined contributions. Developed countries are required to provide financial resources to assist developing countries with mitigation and adaptation efforts, but the amount or method of application of these resources is not specified.

Some in the scientific community were, like the big environmental groups, positive. "The fact that we got an agreement with a temperature target, with a commitment to a direction of travel, with a commitment to improving and enhancing the financing that is going to be necessary to meet that direction, I'm pretty optimistic about it," said Nigel Arnell of the UK's University of Reading.[10] But commitments and aspirations aside, others were not enamoured with the deal. In a letter to the British newspaper *The Independent*, eleven top climate scientists claimed the deal is far too weak to help prevent devastating harm to the Earth.[11]

James Hansen was, if possible, even more condemning. If the world had heeded his warning when he first raised the alarm in 1988, containing global warming within two degrees Celsius would not have been such a daunting task. He called the Paris deal "a fraud really, a fake. . . . It's just worthless words. There is no action, just promises. As long as fossil fuels appear to be the cheapest fuels out there, they will continue to be burned."[12]

What was accomplished in Paris in 2015 feels "like an ambitious agreement designed for about 1995 when the first conference of parties took place in Berlin," writes Bill McKibben, environmentalist and co-founder of 350.org. "If the agreement had been adopted . . . in 1995 [it] might have worked. Even then it wouldn't have completely stopped global warming, but it would have given us a chance of meeting the 1.5 degree Celsius target that the world notionally agreed on."[13]

And Canadian climatologist Gordon McBean, president of the prestigious International Council for Science, said it's "largely not possible" that the world will meet the two degree target. The planet has already warmed about one degree from the pre-industrial level of 1850. Global temperatures have risen about 0.2 degrees per decade for the past several decades, so they will hit 1.5 degrees by 2050 regardless of emissions reductions "because of the time lag of our climate system."[14]

Ecological economist Clive Spash presents a harsher assessment:

> *The Paris Agreement . . . is a fantasy which lacks any actual plan of how to achieve the targets for emissions reduction. There are no mentions of greenhouse gas sources, not a single comment on fossil fuel use, nothing about how to stop the expansion of fracking, shale oil or explorations for gas and oil in the Arctic and Antarctic. Similarly there are no means for enforcement.*[15]

The inability to contain global warming within two degrees was an inevitable consequence of the process Maurice Strong set in motion in 1992 by giving the corporate sector an inside track on climate change negotiations. This may explain why the UNFCCC was unable to establish mandatory emission levels that could slow down temperature increases. Business did everything it could to prevent this. Corporations and their trade associations walked the convention halls and held private discussions with nations trying to reduce their consumption of products manufactured by these same corporate actors. As the conference got underway, a group of NGOs led by the Brussels-based Corporate Europe Observatory submitted a petition to Christiana Figueres asking that big polluters be "kicked out" of climate policy-making. They also questioned Figueres's impartiality since she had been climate-change adviser to Latin America's largest energy company, Endesa, before she became the UN's climate chief in 2010. Her spokesperson, Nick Nuttall, said it was "totally unfair" to rebuke her for working for the energy giant. He did acknowledge that she would "constantly try to engage with the fossil fuel industry," and therefore had no intention of banning Big Oil from the proceedings.[16]

At both COP20 and COP22 the same dynamic was at play. At COP20 in Lima, Peru, climate activists presented a petition to the UN negotiations secretariat with 53,000 signatures calling for fossil fuel corporations and their lobbyists to be banned from the talks.[17] And as a result of decisions made in Paris, corporate interests such as ExxonMobil, Chevron, Peabody, BP, Shell and Rio Tinto had unquestioned access to most discussions at COP22 in Marrakesh, Morocco through the observer status of the bodies that represent these companies such as the World Coal Association and Business Roundtable. At Marrakesh, twenty developing nations representing the majority of the world's population asked for a report that would examine how "the United Nations system and other intergovernmental forums . . . identify and minimize the risks of conflicts of interest." But that request was deleted from the meeting's final report. Australia, the US and the EU spoke against any exclusion of fossil fuel lobbyists. No report was ever produced.[18]

HOW BUSINESS GAINED CONTROL

Corporate participation in COP21 and in the conferences and talks leading up to and following it stands in stark contrast with the corporate role in the World Health Organization's Framework Convention on Tobacco Control. There, tobacco interests are excluded, a fact which helps explain that treaty's rapid progress in curtailing tobacco use.[19] The convention was adopted in 2003 after just three years of talks and despite the efforts of the tobacco denial machine. Article 5.3 of the convention says "In setting and implementing their public health policies with respect to tobacco control, Parties shall act to protect these policies from commercial and other vested interests of the tobacco industry in accordance with national law."[20] The World Health Organization even provides a thirteen-page set of guidelines for how to implement article 5.3. The guiding principle is that "there is a fundamental and irreconcilable conflict between the tobacco industry's interests and public health policy interests."[21]

At the climate talks, in sharp contrast, there is no conflict between Big Oil's interests and public health and environmental interests. The corporate sector succeeded in making itself integral to the process. "Business played a major role in making the COP21 a success," Ségolène Royal, the French Minister for Ecology, Sustainable Development and Energy confirmed at a follow-up meeting with business leaders in Paris in February 2016. "They have made a decisive difference compared to previous climate summits."[22] As for those who criticized corporate participation, the UNFCCC's Figueres says it is time to stop "pointing the blaming finger at fossil fuel companies." Their technical expertise and "amazing power" can help to slow the rise of greenhouse gas emissions, she told business leaders in Barcelona in the run-up to the Paris talks. "Bringing them with us has more strength than demonizing them."[23]

Yet it was Big Oil's technical expertise and amazing power that caused greenhouse gas emissions to soar in the first place. Its power was demonstrated by its ability to repeatedly block almost all attempts to bring in stringent environmental regulations over three decades. The longer global businesses wait to start cutting their emissions, the steeper the rate of decline must be in subsequent years.[24] How can one *not* point the blaming finger at these companies? At COP21, in contrast, Big Oil corporations like ExxonMobil, Royal Dutch Shell and BP positioned themselves as strong climate defenders and claimed to have been taking action to reduce greenhouse gas emissions in their various operations for many years, which was an obvious falsehood.[25] It was little more than an expensive public relations exercise designed to embed its members in UNFCCC inner circles.

To further its objectives, Big Oil even partly financed the talks. France could have easily paid the CAD$255 million cost, but by allowing corporations to contribute 20 per cent, the host country encouraged the private sector to be part of the inner circle that was planning and organizing the event. One such sponsor was Engie, Europe's largest natural gas and electricity provider, with 2017 revenues of US$71 billion. Engie is a fabled French company, one-third owned by the French government, with roots going back to the Universal Suez Canal Company, founded in 1858 to build the Suez Canal. But that was then. Today it is the world's eighth-largest emitter of greenhouse gases. It says it is moving to hydropower and renewable energy, but ten of its coal-fired power plants are considered "subcritical," making it the third-least-carbon-efficient company in the entire energy sector. Subcritical is defined as "the least efficient and most polluting form of coal-fired generation" in a 2015 report on stranded assets and subcritical coal.[26] "Sub-critical coal-fired power stations are the first thing we need to kill off if we want to tackle climate change," explains Ben Caldecott, the Oxford University researcher who was lead author of the report that evaluated the carbon emissions of large energy companies with coals assets.[27]

The company worked hard to clean up its image before the summit, to move from being seen as part of the problem to being part of the solution. It changed its name from stodgy GDF Suez to Engie. This is supposed to make hearers think of energy transition rather than dirty coal. With the new name came a new logo — the rising sun, symbolizing not Japan but "a new day in the world of energy."[28] Closer to the conference, it launched a major rebranding exercise in which Engie is identified as a leader in the energy transition and an indispensable renewable energy player.[29]

As for actual change, not too much. The company did promise to stop investing in coal, a move provoked in large part by a vigorous campaign by Friends of the Earth France to target Engie's "dirty energy investments."[30] This would hardly be a good fit for a company sponsoring the climate talks. At the same time though, Engie was heavily invested in research into carbon capture and storage, which would allow coal-fired power plants (and

Canadian oil sands plants) to continue operating as long as they can bury or reuse their carbon dioxide emissions.[31]

Despite Engie's dirty-polluter profile, the company seems to be a favourite of the French government, perhaps because it's the dominant owner. In a seeming pay-to-play scheme, COP president Laurent Fabius asked Engie CEO Gérard Mestrallet to co-chair a "Business Dialogue" with France's special representative for climate change, Laurence Tubiana. This effort brought together thirty CEOs with an equal number of negotiators. It was "a large, structured mobilization of companies," Mestrallet admitted.[32] Their four meetings focused on two issues: how to get carbon pricing into the agreement when it doesn't fall under the mandate of UN climate negotiations, and how to expand the market for low-carbon technologies and products. It was significant because business "was able to propose and introduce into the Paris agreement the principle of a carbon price sign," Mestrallet said. The business dialogue format allowed industry to execute an end run around the COP21 structure.

ENTRENCHING EMISSIONS TRADING IN THE AGREEMENT

Business wanted a price on carbon because that becomes another cost of doing business it can factor into its strategic planning. Six months before Paris, major European-based energy companies — BP, Shell, Statoil, Total and others — wrote to Figueres and Fabius, calling for carbon pricing in the agreement.[33] At the same time, business didn't want binding targets enshrined in the agreement because that would impede its growth and profitability. It got both at COP21. The phrases "emissions trading" and "carbon price sign" don't appear in the Paris text, but that's what is meant in Article 6, which lays out rules for countries that choose to engage "on a voluntary basis in cooperative approaches that involve the use of internationally transferred mitigation outcomes towards nationally determined contributions."[34] In plain English, this article authorizes countries to participate in carbon markets as a means of reducing greenhouse gas emissions by buying credits from other nations.

Two government representatives, Catherine McKenna of Canada, and The Democratic Republic of Congo's Foreign Affairs Minister, Raymond Tshibanda N'Tungamulongo, led an informal consultation that put forward proposals presented to the UNFCCC by the International Emissions Trading Association. (Canada may have taken the lead in this effort because it already knew it could never meet its "nationally determined contribution" without buying credits from other countries.) Careful wording was necessary to avoid the objections of large fossil fuel exporters like Saudi Arabia and developing countries like Bolivia. When countries choose to participate in carbon markets, the emission reductions are correctly tracked and accounted for. This is what the International Emissions Trading Association wanted — and got.[35]

The association is a global organization of oil and gas companies, banks, law firms and consultancies organized by BP that promotes carbon markets. With over a hundred delegates each at COP21, the International Emissions Trading Association and the World Business Council for Sustainable Development (WBCSD) sponsored a business hub within the conference's blue zone, where access was restricted to official delegates. The hub was framed as the "home" for business to voice its priorities and solutions; about a hundred side events were held there.

Business had worked long and hard to heal the divisions that undermined its effectiveness at earlier conferences. Front and centre was the difference in approach between European energy companies BP and Shell and the American ExxonMobil. Corporations prepared for COP21 well in advance of the formal meetings through a series of events, enabling business associations and their members to build a unified "business voice."[36] Paul Polman, chair of the WBCSD and CEO of food and cosmetics giant Unilever, was everywhere. He spoke at many Paris events and his company was one of the most active. Polman was "leader" of the business organizations We Mean Business and the B-Team, a board member of the UN Global Compact, the International Chamber of Commerce and the Consumer Goods Forum, as well as the World Economic Forum International Business Council and the Global Commission on the Economy and Climate, creating a vast network of corporate connections.[37]

In its summary of conference proceedings, the International Emissions Trading Association bragged that all three of the organization's pre-COP asks were included in Article 6 of the agreement, although not in these words: the right of countries to trade emission-reduction units; rules for carbon market accounting; and a new crediting mechanism.[38] The idea of a Carbon Pricing Leadership Coalition was initiated by the World Bank Group in September 2014 and officially launched at COP21. Support was gathered at UN Climate Week from seventy-four countries and more than 1,000 companies. The momentum was unstoppable.[39] Even Justin Trudeau highlighted carbon pricing in his head-of-state remarks delivered on Day One of the conference. And it wasn't just the International Emissions Trading Association and Trudeau. The Environmental Defense Fund bragged on its website about how it pushed in the corridors of Paris for an "opening for markets."

At the same time business was pushing markets into the agreement, it was avoiding binding targets. This success was revealed in a last-minute dispute over language in a key clause. In non-legalese language, the final version of article 4.4 of the agreement says that developed countries "*should* have economy-wide absolute emission reduction targets" (italics added).[40] The draft version used the word "shall" instead of "should," meaning a country must have absolute emission-reduction targets enshrined in legislation. Countries such as South Africa and Nicaragua wanted the more binding word kept in. The US under Democratic President Barack Obama could never accept this because a Republican Congress would never pass such a law — its fossil fuel backers would never allow it. Thanks to the negotiating and persuasion skills of French Foreign Minister Fabius — he called it a "typing error" — Big Oil and its surrogate, the US government, won.[41]

Big Oil's victory was even deeper. Even though the production and use of fossil fuels are responsible for 70 per cent of global emissions, fossil fuels are not mentioned in the agreement. Both the 1992 Framework Convention and the 1997 Kyoto Protocol mention energy questions, but these are notably absent from Paris. In a concerted effort at linguistic cleansing, Big Oil and the major oil-producing nations, led by Saudi Arabia, were successful in keeping energy issues out of the treaty. The word "energy" is used three times: once to refer to the Atomic Energy Agency, once in the phrase "sustainable energy," and the third in the phrase "renewable energy." Fossil fuel energy is nowhere to be found. In negotiating documents produced before Paris, in contrast, one clause mentions the "phasing down of high-carbon investments and fossil fuel subsidies." Other passages refer to

"a low-carbon transition," "a zero emissions" target and "full decarbonisa-
tion." All gone from the final document. Instead it is filled with intentions
and aspirations.[42]

Big Business seemed proud of its achievements at COP21. Paul Polman
reflected the self-congratulatory mood of the CEOs by saying that "achieving a zero emissions economy is the greatest business opportunity of the
century." Polman tied climate change mitigation to economic growth:
"if we don't tackle climate change, we won't sustain economic growth or
end poverty." Such a realization "will unlock trillions of dollars and the
immense creativity and innovation of the private sector who will rise to the
challenge in a way that will avert the worst effects of climate change," he
declared.[43]

The only legally binding requirement in the agreement is that parties
must report their emissions regularly.[44] They are urged to strengthen their
commitments every five years, beginning in 2020, but this is not legally
required. They are not penalized for missing their targets, targets which in
total do not meet the two-degree limit. If all parties keep their promises, the
planet would still warm by 3.5 degrees Celsius above pre-industrial levels.
And waiting five years from the approval date of 2015 until a new carbon
budget is calculated means five more years of carbon dioxide accumulation. Carbon cuts needed to start immediately if there was to be any hope
of keeping the temperature increase to 1.5 degree. But that didn't happen.

The rhetoric of sustainable development, launched by Maurice Strong
and Stephan Schmidheiny at Rio in 1992, underlies the agreement. Business
is comfortable with the use of that term since it gained control over it in the
1987 Brundtland Commission report. It brought the concept into the global warming process through the World Business Council for Sustainable
Development, but they meant sustainable corporate development; for a
policy to be acceptable to business it must sustain corporate growth and
profits. The phrase appears sixteen times in the Paris Agreement.

THINK TANKS CONTINUE TO DENY

While the World Business Council for Sustainable Development and the
International Emissions Trading Association dominated business influence
at COP21, the think tank wing of the assault on environmentalism was

also present, albeit with reduced impact. But business continued to support the denial efforts of neoliberal think tanks. Creating doubt and confusion about climate change was always a good investment. For the first twenty years, most funding came from ExxonMobil, other Big Oil companies and the foundations of a dozen wealthy conservative business executives. Donations from these sources tailed off after 2008, however, while donations from Donors Trust and Donors Capital skyrocketed, to become the central component in funding global warming denial. Donors Trust and Donors Capital are donor-directed foundations in which individuals, corporations or other foundations make contributions to them and they then make grants based on the preferences of the original donor without revealing the donor's identity. "Dark money," as it came to be called, took over the funding of global warming denial.[45]

Two think tanks that led the early counter-movement in the 1990s maintained denial leadership over the ensuing years, as they continued to dispute the accuracy of climate science and question the integrity of climate scientists.

The Heartland Institute focuses on the uncertainty of global warming science as the science becomes ever more certain. A study of nearly 12,000 peer-reviewed scientific papers indicates that of the papers expressing a position on anthropogenic climate change, 97 per cent endorse the consensus position that humans are causing global warming.[46] The institute will have none of this. It was founded in 1984 by Chicago investment banker David Padden, who was a member of the Mont Pèlerin Society and was closely affiliated with the Koch brothers.[47] The Economist calls Heartland "the world's most prominent think tank promoting skepticism about man-made climate change,"[48] while the New York Times says it's "the primary American organization pushing climate change skepticism."[49] To earn these accolades, Heartland organizes an annual international conference on climate change that serves as a forum for deniers. The think tank made worldwide headlines in 2012 after it launched an ad campaign associating acceptance of climate science with "murderers, tyrants and madmen" that include Ted Kaczynski (the Unabomber), Charles Manson and Fidel Castro. It continues to soldier on, with declining impact.[50] The institute held its annual conference in Paris in conjunction with the official conference, attracting an audience of thirty.[51]

THE BIG STALL

The other leading house of denial has been the Competitive Enterprise Institute. Political writer and lobbyist Fred L. Smith worked for the Koch-funded Council for a Competitive Economy before founding this institute in Washington, D.C. in 1984, the same year as Heartland, with Koch brothers support. The Competitive Enterprise Institute fought against tobacco regulation before it turned its guns on "alarmist views" of global warming. Smith wasn't a member of the Mont Pèlerin Society, but he did give presentations at some of the society's gatherings.[52] He was an ardent advocate for free-market environmentalism, the belief that the environment can be protected only if it is fully privatized — every tree, lake, river, even air shed.[53]

Leadership on the denial file passed from Smith to Myron Ebell, the Competitive Enterprise Institute's director of energy and environment. Ebell excelled in ferocious denial advocacy. He's considered enemy number one by the climate-change community.[54]

So it may have come as little surprise when presidential candidate and denier Donald Trump selected Ebell to lead the transition team for the Environmental Protection Agency, the organization most responsible for governmental responses to global warming. The team selected Oklahoma Attorney General Scott Pruitt, who had sued the Environmental Protection Agency fourteen times regarding agency actions and who told the Senate Environment and Public Works Committee at his confirmation hearing that, "science tells us the climate is changing and human activity in some manner impacts that change. The ability to measure the precision, degree and extent of the impact and what to do about that are subject to continued debate and dialogue."[55]

In 2018 denial continues to work, at least in the US. A 2016 Pew Research Center poll found that only 50 per cent of American registered voters and just 22 per cent of Trump supporters said that climate change is mostly due to human activity.[56] For those who hoped the denial machine had finally moved on from attacking climate science to a more fruitful discussion over appropriate climate policies, political scientists Constantine Boussalis and Travis Coan had bad news: "The era of climate science denial is not over."[57]

TRUMP OPTS OUT, BUSINESS DOESN'T

Just ask Donald Trump, when he pulled the US out of the Paris Agreement. He garnered support for his action in a letter signed by thirty-eight neo-liberal think tanks, AstroTurf organizations and Tea Party groups.[58] The letter was led by the Competitive Enterprise Institute, Heartland and the American Energy Alliance. The latter is a lobbying and advocacy group for coal and gas companies that has received significant financial support from Koch Industries. Alliance president Thomas Pyle was a former Koch and oil industry lobbyist who headed Trump's transition team for the Energy Department, successfully placing denialists Scott Pruitt as head of the Environmental Protection Agency and former Texas Governor Rick Perry as Secretary of Energy.

The letter, which was posted on the American Energy Alliance website, concluded: "The undersigned organizations believe that withdrawing completely from Paris is a key part of your plan to protect US energy producers and manufacturers from regulatory warfare not just for the next four years but also for decades to come." These groups didn't have to concern themselves about how Trump's withdrawal might negatively impact efforts to contain and decrease greenhouse gas emissions because, of course, they don't believe such emissions lead to rising temperatures, or say they don't believe it. "The Paris Climate Treaty is an all-pain-for-no-gain agreement that will produce no measurable climate benefits and exacerbate energy poverty around the globe," the Competitive Enterprise Institute's Myron Ebell said.[59] Trump senior adviser Steve Bannon and Scott Pruitt both urged Trump to exit.

But actual businesses, at least those prepared to go on the record, urged Trump to stay in the game. Twenty-five major companies, including brands such as Apple, Facebook, Gap, Hewlett Packard, Levi Strauss, Mars, Microsoft and Morgan Stanley, took out full-page ads in *The New York Times* and *Wall Street Journal* urging the president "to keep the United States in the Paris Agreement." The ad claimed that staying in would strengthen competitiveness, create jobs, markets and growth and reduce business risks.[60] These companies were joined by Trump daughter Ivanka and son-in-law Jared Kushner who urged Trump to stay in.

Bannon and Pruitt won. After Trump made his announcement — it "hamstrings the United States" and would cost jobs in industries like coal and steel, he said — some business executives responded angrily. "Climate change is real," Jeff Immelt, CEO of General Electric, said in a tweet. "Industry must now lead and not depend on government." Elon Musk, head of electric carmaker Tesla, resigned from the president's business advisory council. "Leaving Paris is not good for America," he tweeted.[61]

One of the most crucial bits of business input came several months earlier, in a letter to the president from ExxonMobil. The company told the White House it believed the agreement was an "effective framework for addressing the risks of climate change" and the US should remain a party to it. ExxonMobil CEO Darren Woods, who replaced Rex Tillerson as head of the company when Tillerson left to become Trump's secretary of state, wrote a blog post saying the company is "encouraged" that the Paris agreement creates a framework for all countries to address rising emissions.[62]

It was quite a change for the global giant, which was being investigated for allegedly misleading shareholders and the public regarding what it knew about the dangers of climate change and when it knew it. ExxonMobil was a major backer of the Global Climate Coalition and one of the largest contributors to the denial machine — at least until 2008 when, like most of big business, it realized that a predictable future and an "effective framework" were preferable to ongoing uncertainty about climate change policy. In a second letter to Trump, Woods argued that staying in the agreement would ensure the "United States will maintain a seat at the negotiating table." At the company's annual general meeting in May 2017, he said staying in was vital "to ensure a level playing field so that all energy sources and technologies are treated equally."[63]

THERE'S ALWAYS GEOENGINEERING

The Paris Agreement may be what Big Oil wants, but the agreement will never get the world anywhere near where it needs to be to keep climate change within tolerable limits. The fossil fuel industry signed on for carbon pricing and trading, but this strategic change of direction was not intended to keep global warming within two degrees. Rather, the purpose of emissions trading and carbon taxes was to ensure markets continue to operate. Eventually it will become clear that pricing carbon, investing in clean technology and creeping toward a low-carbon future will not do the job. Then politicians, policy-makers and financiers will begin to entertain geoengineering solutions — major technical and engineering disruptions of the climate system. They already are.

Big Oil embraced denial. Then it embraced carbon pricing. As of 2018, it is considering geoengineering. These three efforts to deal with global warming are not unrelated or rival panaceas, explains Notre Dame University's Philip Mirowski, but "together constitute the full-spectrum neoliberal response to the challenge of global warming." They are neoliberal because they arose from the network of think tanks and academic units affiliated with the Mont Pèlerin Society. They are neoliberal because they are designed to ensure the market, and not the state, deals with the problem.[64]

Promoting denialism bought time for markets to be created. They were enshrined in the Paris Agreement. After they are seen to be failing, entrepreneurs with patents for atmospheric manipulation and financiers with deep pockets are lining up to obtain official approval to move ahead and make lots of money.[65] For neoliberals, the market must continue even though most species and a sizeable chunk of humanity may not.

Further evidence for the neoliberal roots of these efforts is provided by the fact that some neoliberal think tanks that challenged established climate science are now promoting geoengineering. They include the Hoover Institution, Heartland Institute and American Enterprise Institute. Why would they push geoengineering solutions to climate change if climate change wasn't occurring, one might ask. The answer: to prevent the United Nations and national governments from taking charge.

Geoengineering strategies fall into two categories: removing carbon dioxide from the atmosphere — carbon capture and storage — and reducing heat by reflecting sunlight back into space. The former is being done but is expensive unless taxpayers contribute to it; the latter is not yet being done, but may be "wildly, utterly, howlingly barking mad," according to Oxford University physics professor Raymond Pierrehumbert in a 2015 article in *Slate*.[66] It's like "jumping off the Washington Monument and hoping someone invents antigravity before you hit the ground," Pierrehumbert, a lead author in the Intergovernmental Panel on Climate Change's Third

Assessment Report, said in an interview.[67] He is referring specifically to solar radiation management, in which the sun's rays are reflected back into space.

A leader in the charge to develop solar radiation management is Harvard University physics professor David Keith. He recognizes the dangers but believes solar radiation management needs to be thoroughly researched. Keith is a favourite of Bill Gates, who has invested millions of dollars in Keith's research.

Solar radiation management is meant to mimic volcanic eruptions that send large amounts of sulphur dioxide into the stratosphere, triggering temporary cooling events. For example, the eruption of Mount Pinatubo in the Philippines in 1991 sent about 17 million tonnes of sulphur dioxide into the atmosphere and cooled the northern hemisphere by 0.5 to 0.6 degrees Celsius.

One proposal is to use a fleet of modified business jets to inject fine droplets of sulphuric acid into the stratosphere, where they will combine with water vapour to form fine sulphate particles that reflect sunlight away from the planet. One study estimates that to achieve one degree of cooling will require 6,700 flights a day of high-flying jets carrying tanks of sulphur and would cost US$20 billion a year.[68]

Solar radiation management raises numerous issues. At its core, it adds one pollutant to counteract another. It could deplete the ozone layer, harm plant and animal species, cause air-pollution deaths, and stir global conflict by benefitting some countries while causing droughts in others. Sulphuric acid-dispensing flights would have to go on forever or at least until humanity reduced the concentration of greenhouse gas emissions in the atmosphere. And that might not happen because solar radiation management will likely be perceived as a "get-out-of-jail-free card." People will continue to burn fossil fuels secure in the knowledge they have a fallback. Politicians will realize they need not worry about reforming the energy system.[69]

Carbon capture and storage, the other type of geoengineering, has been in use for a few years. Canada has two operating plants: Shell's Quest facility as part of the Athabasca Oil Sands Project and SaskPower's capture-and-storage units at the Boundary Dam Power Station near Estevan, Saskatchewan. They capture carbon dioxide from upgrading bitumen or burning coal before it is emitted into the atmosphere and bury it deep underground or use it in other industrial processes. About a dozen projects are operating around the world. Help is on the way for them at least in the US. Donald Trump's February 2018 budget includes tax credits of up to US$50 a tonne for captured carbon dioxide, a move that is likely to encourage the construction of more capture-and-storage plants to sequester carbon dioxide created by burning coal.

But what about the carbon dioxide already in the atmosphere? With Bill Gates's funding, David Keith has formed a for-profit company to capture carbon from the atmosphere. At a facility in Squamish, B.C., Keith has developed a patented technology that can capture carbon dioxide from the air and turn it into a low-carbon transportation fuel that can replace traditional gasoline at a competitive price.[70] If the technology works it's likely to be a profitable venture for inventor and investor, but will do little to address the 36 billion tonnes of greenhouse gas emitted into the atmosphere every year.

CHAPTER 11
OIL SANDS BINGES, REBRANDS

On June 3, 1996, Canada's Declaration of Opportunity for the oil sands was signed in Fort McMurray, Alberta. Prime Minister Jean Chrétien signed for the federal government, Alberta energy minister Pat Black for the province. It was just one of many signs that the stars were aligning for a vast scale-up in oil sands production. The incentives of federal and provincial government financial support and significant cost reductions due to improvements in mining and drilling techniques spurred domestic and foreign companies to invest in the oil sands. The immense value of the resource became evident to everyone in the industry.

It soon became clear that Chrétien and the Liberals were not intending to impose any major reductions in greenhouse gas emissions despite their support for Kyoto. The voluntary approach would be good enough for them.

The rapid run-up in the price of oil was the icing on the cake. Oil moved above US$20 a barrel in October 1999 for the first time since the mid-1980s and briefly hit US$140 a barrel in 2008. The Klein government's hands-off policy — let the market decide which projects go forward — stood in stark contrast to Peter Lougheed's careful consideration on a project-by-project basis in the 1970s. At the height of the boom, Lougheed was so worried he called for a hiatus, but Klein and his successor, Ed Stelmach, paid him little heed.[1]

Although the private sector benefited the most from rock-bottom royalty rates and generous tax breaks, it was the public sector — or at least its pale Alberta version — that was instrumental in boosting industry's prospects. A key player in oil sands development was Alberta Energy Corporation, the company started by Lougheed to ensure Albertans could participate in their province's oil and gas development. This initiative was undone by Klein when he privatized the company.

But Alberta Energy Corporation had been experimenting with a new technique called steam-assisted gravity drainage to unlock bitumen too deep to be mined, or in-situ production. This was no trivial matter, since about 80 per cent of bitumen deposits couldn't be extracted through mining.

Steam-assisted gravity drainage was invented by an Imperial Oil engineer in the 1970s, but was developed by the Alberta Energy Research Institute, a Crown corporation created by Lougheed to develop new technologies for oil sands and heavy oil production. The institute set up a test facility near the Syncrude and Suncor mining operations north of Fort McMurray, and developed the new technique on a cost-shared basis with industry.[2]

Alberta Energy Corporation's property, Foster Creek, 250 kilometres northeast of Edmonton, was one of the first to apply the technique commercially, producing its first oil in 2001. It was followed a year later by the Mackay River project, a second in-situ plant that used steam-assisted gravity drainage, developed by Petro-Canada, another company with public-sector roots. After that, the stampede was on, as both foreign companies — Devon Canada, ConocoPhillips Canada, Husky Energy, and Nexen Energy — and domestic companies — Petro-Canada, Suncor Energy, Canadian Natural Resources — pushed projects forward. And along with the vast scale-up in production came the inevitable environmental and social impacts: greenhouse gas emissions, excess water usage, boreal forest destruction, pressures on wildlife, negative impacts on First Nations and housing shortages. But this didn't seem to deter any project from moving forward. Production nearly tripled, from 445,000 barrels a day in 1996, when Chrétien and Black signed the Declaration of Opportunity, to 1.2 million barrels a day in

2008 when oil prices peaked. And production continued to soar, doubling to 2.4 million barrels a day in 2016.[3]

The Canadian subsidiaries of Seven Sisters giants Royal Dutch Shell and Exxon had been in Canada almost since the beginning of the oil age. But Shell Canada and Imperial Oil took different routes to exploiting the oil sands. As early as 1978, the year Syncrude came on line, Shell announced plans for a mining operation. Two decades would pass — and the tax and royalty inducements put in place — before Shell gave the green light to proceed with the third large mining operation, Athabasca Oil Sands Project. Shell acquired one of the prime leases with vast deposits in the late 1950s, but each time it put together a proposal it could foresee profits only in the single digits.[4] Now prospects were rosier, thanks to higher oil prices, and Shell began production in 2002.

Imperial Oil was one of the original participants in the Syncrude project, and has remained a partner for fifty years. Even before Syncrude, though, Imperial began experimenting with techniques to extract bitumen from the oil sands too deep to be mined. It had acquired leases in the Cold Lake area northeast of Edmonton with vast reserves located about 500 metres below ground. Imperial perfected an in-situ recovery technology called cyclic steam stimulation. This technique involves putting a well through cycles of steam injection, soaking and pumping. The project opened for business in 1985, and was producing 90,000 barrels of oil equivalent per day when Chrétien and Black signed their declaration.[5]

During the first decade of the twenty-first century, foreign companies flooded into the province, spending CAD$30 billion buying oil sands companies, assets and leases. There were good reasons for an influx of foreign capital. The world seemed to be running out of oil and here was the province of Alberta sitting on the third-largest proven bitumen reserves in the world. Yes, it was costly to extract and upgrade into a usable product, but oil prices had been near triple-digit territory for most of the last decade. Who wouldn't want a piece of the action, especially given the accommodating and permissive policies of the federal and Alberta governments? Foreign investors continued to buy shares in oil sands companies and foreign firms continued to buy up Canadian-based ones.

Statoil of Norway in 2007 paid US$2.2 billion to buy out North

American Oil Sands, a private company with major leases in the Athabasca region south of Fort McMurray. Total of France purchased several companies that were planning to develop mining and steam-assisted gravity drainage projects, spending US$3.6 billion. But it was the US$15 billion acquisition of Calgary-based Nexen Energy by state-owned China National Offshore Oil Corporation that put foreign ownership on the Harper government agenda. Foreign ownership of Canadian resources was fine and to be encouraged, Harper's natural resources minister, Joe Oliver, explained, but "what we were concerned about was the ownership by a government."[6] Harper allowed the deal to go through, but his government introduced new rules to limit state-owned companies from buying oil sands firms, emphasizing they'd be approved only in "exceptional" circumstances.

PIPELINES UP THE ANTE

Even before most oil sands developers put a shovel in the ground, pipeline companies were considering their options. Three major companies — Enbridge, TransCanada Corporation and Kinder Morgan — jockeyed to be first, or at least second, to cash in on the vast resource by providing additional transportation capacity.

Enbridge is the largest oil pipeline operator in Canada as of 2018. Its Mainline system — from Edmonton to Superior, Wisconsin — is as old as the modern Canadian oil industry, construction having begun in 1950, just three years after oil was discovered near Leduc. Together with the Lakehead system, Enbridge has been the main conduit for Alberta and Saskatchewan production to North American markets. Over recent decades, light crude has given way to heavier diluted bitumen. In 2016, Enbridge's production ramped up to 2.8 million barrels per day of various oil products, accounting for nearly two-thirds of exports to the US. Some product travels east to refineries in Montreal, but most turns south in Wisconsin.

While the environmental movement fixated on Enbridge's CAD$7.9 billion Northern Gateway proposal to bring 525,000 barrels per day of diluted bitumen to the west coast for transport to foreign markets, Enbridge quietly forged ahead with major increases in its capacity to export product three times the volume of Northern Gateway. The first move occurred in 1999, when Enbridge's Athabasca pipeline connected Alberta's oil sands projects to its mainline system in Hardisty, Alberta. By 2014, twelve oil sands projects were connected to the Enbridge system.

To get this oil to market, Enbridge pushed forward on multiple fronts. It replaced the entire pipeline from Hardisty to Superior, Wisconsin, with larger pipe and more pumping stations. It extended the line to refineries in eastern Canada, including the reversal of its line that ran from Montreal to Sarnia. The most significant move was a vast expansion of the pipeline system that ran from Illinois to the large refinery complex on the western Gulf Coast near Houston, Texas.

In Canada, the company's major expansion was the replacement of Line 3 that ran to Wisconsin. It also bought existing pipelines, a more certain option for increasing capacity than going through the lengthy approval and construction process. And it pushed forward on the natural gas front with its blockbuster takeover of Houston-based Spectra Energy Corporation for CAD$37 billion. Spectra's business centred on natural gas pipelines accessing shale gas formations in the northeastern United States. Enbridge was positioning itself to benefit from Big Oil's game plan to frame natural gas as the poster child of the low-carbon future, even though natural gas is not low-carbon. The takeover created North America's largest energy infrastructure company.

In 2005, Houston-based Kinder Morgan bid US$5.6 billion for Terasen, operator of the Trans Mountain Pipeline, which carries 300,000 barrels per day of heavy oil and diluted bitumen from Edmonton to Vancouver and Puget Sound. A solid connection to Alberta's booming oil sands was the driving force, Kinder Morgan's billionaire CEO Richard Kinder explained when announcing the bid. Terasen, the former BC Gas and before that the privately owned Inland Natural Gas, already had plans to twin the pipeline. Ten years later, its CAD$7.4 billion project to twin the line and boost transmission to 890,000 barrels per day remained mired in controversy.

Like Kinder Morgan, TransCanada Corporation was a late entrant into the Canadian heavy oil pipeline business. For decades it prospered by methodically covering North America with pipelines connecting gas fields with markets. It also became the largest private-sector electricity producer, finding a use for some of that gas. At the same time, it was part of a consortium that operated the Bruce Nuclear Generating Station that provided about 30 per cent of Ontario's electricity.

The same year that Kinder Morgan snapped up Terasen, TransCanada Corporation decided the time was right to get into the diluted bitumen transmission business, with its original Keystone pipeline proposal. The gas transmission business, while still healthy, was stagnating compared to diluted bitumen. Keystone was completed in 2010 without controversy, transporting 435,000 barrels per day of synthetic crude and diluted bitumen from its terminal in Hardisty, Alberta, to the junction at Steele City, Nebraska and then east to refineries and a terminal in Illinois. The line was extended south to the storage and distribution centre in Cushing, Oklahoma and from there to refineries at Port Arthur, Texas on the Gulf Coast. New pumping stations increased the pipeline's capacity to 591,000 barrels per day. Keystone was the centrepiece of the company's strategy to ship bitumen to the Gulf Coast.

In 2008, with no end in sight to higher oil prices, even before Keystone was completed, TransCanada proposed a major extension of the system, the notorious Keystone XL (export limited). KXL would duplicate the Phase 1 pipeline, but over a shorter route and with larger diameter pipe, giving it the capacity to transport up to 830,000 barrels per day. The KXL would connect TransCanada's terminal at Hardisty directly to Steele City, Nebraska, where it would connect to the existing Keystone system.

TransCanada ran into stiff and growing opposition to the pipeline, which would traverse Montana, South Dakota and Nebraska. The project became hopelessly bogged down in American and Canadian politics. As well, the US was beginning to require fewer imports due to the explosion of gas and oil fracking. Despite the millions of dollars TransCanada and oil industry advocacy groups like the Consumer Energy Alliance paid to promote the pipeline, President Barack Obama rejected the application.[7] It was dead, at least until Donald Trump was installed as president and resurrected the project.

In 2013, with KXL going nowhere, TransCanada came up with its Energy East pipeline proposal to ship crude oil to the east coast of Canada. This pipeline was first proposed by Derek Burney in a *Globe and Mail* op-ed. Burney was well-placed to help navigate the project. He joined the TransCanada board of directors in 2005, when the company decided to get into the business of exporting heavy oil and diluted bitumen to the US.

He had been Canada's ambassador to the US under Brian Mulroney and a central player in negotiating the Canada-US and North American free trade agreements — with their proportionality provisions that prevented Canada from protecting its oil resources for use by Canadian consumers.

In 2011, when KXL and Enbridge's Northern Gateway appeared to be in trouble, he proposed an oil pipeline to the east coast. From there the product would go by ship to Asia, bypassing opposition and regulatory roadblocks for pipelines aiming for the US and the west coast.[8] Two years later the company submitted its CAD$15.7 billion Energy East proposal to the National Energy Board. For much of the route it would use underutilized natural gas pipeline. While the company inched forward with its oil pipeline projects, it continued to expand its gas operations with the US$10.2 billion purchase of Columbia Pipeline Group, which operates largely in the massive shale region of the northeastern United States, further reinforcing its role in natural gas transmission. Finally, in October 2017, despite the best efforts of the company, its lobbyists and the Alberta government, the company cancelled Energy East. With Keystone back on the table, the business case for an additional 1.1 million-barrels-a-day pipeline to the east coast collapsed.

THE CRASH AND THE COMING OF RACHEL NOTLEY

By 2014, each major had one or more projects in the race. Kinder Morgan's Trans Mountain pipeline would be approved by the Trudeau government but would then run into significant opposition from the B.C. government, First Nations and environmental groups. As we've seen, TransCanada's KXL would be rejected by the Obama administration, only to be revived by Donald Trump, although it would still face opposition in Nebraska. Enbridge's Line 3 replacement would also be approved by Trudeau but it still required Minnesota Public Utilities Commission approval.

Given oil sands expansion plans, only two pipelines would be needed. On the oil sands front, foreign and domestic capital pushed on with its projects. Oil prices hovered between US$80 and $100 a barrel after the market resurgence following the 2008 financial collapse.

Then a double blow hit the industry, one economic, the other political. Prices dropped again, falling from $100 to $40 in a matter of months, a price at which it would be more difficult to justify forging ahead with multi-billion-dollar investments. The first sign conditions were going south occurred when French giant Total shelved plans for a new CAD$11 billion mine after prices slipped below $100. It had already cancelled plans to build an CAD$11.6 billion upgrader.[9] The decline in prices led to revenue crashes for everyone. Imperial Oil's revenues were on the way down from CAD$36.2 billion in 2014 to $26.8 billion in 2015, a drop of 25 per cent.

As the May 2015 Alberta provincial election approached, NDP leader Rachel Notley faced a daunting task. Not only did she have to counter the prevailing wisdom in corporate media and boardrooms that the NDP was anti-business to its core, — she also had to face an opponent, Jim Prentice, who was Big Oil's favourite son. A 2014 fundraising dinner Prentice held to support his bid to become leader of the Alberta Progressive Conservative Party, and therefore premier, was sponsored by some of the industry's biggest names: Imperial Oil, Cenovus Energy, ARC Energy Resources, TransCanada Corporation, MEG Energy Corporation and Penn West. They were confident he could end the roadblocks and get some pipelines built.[10]

Events unfolded somewhat differently than advertised. Rachel Notley and the New Democrats swept into power. Albertans "voted for change, and that change includes the environmental record of the province," Ed Whittingham, president of Calgary-based clean-energy think tank Pembina Institute, said. "First and foremost, that means dealing with greenhouse gas emissions."[11]

Most oil executives demonstrated patience and gave the new government time to find its footing. But not Murray Edwards and Canadian Natural Resources. Less than three weeks after the election, the company cancelled its investor day, arguing it couldn't allocate cash until the NDP clarified its position on royalties, taxes, environmental policies and greenhouse gases.[12] Then the company blamed the Notley government for a deferred income tax liability of $579 million because it was increasing the corporate tax rate from 10 to 12 per cent. "This change effectively translates into lower future cash flows and therefore lowers reinvestment in the business," the company said.[13] Notley was face to face with the personification of oil sands power.

Such hardball tactics had made Edwards Canada's thirty-fourth wealthiest person, with a 2018 net worth of CAD$2.88 billion. His fortune was centred on his ownership of 21.3 million shares of Canadian Natural (worth CAD$825 million in March, 2018); he also has holdings in other energy companies.

And with 36 per cent of the shares, Edwards controlled Imperial Metals Corporation, which owned and operated the Mount Polley gold and copper mine near Quesnel Lake in central B.C. The mine's tailings dam collapsed in August 2014, spewing millions of litres of toxic waste into nearby streams, rivers and lakes — and drawing unwanted attention to the low-profile Edwards. He also lost CAD$185 million overnight when Imperial's shares dropped following the mine's collapse.

Several investigations into the circumstances of the dam's catastrophic failure were launched. But no charges were ever laid.

Jaded observers noted that Edwards was a good friend to the Christy Clark Liberal government in B.C. Through his various holdings, Edwards contributed CAD$343,000 to the Liberals during the 2013 election cycle. He also hosted two fundraising events: a private dinner at Calgary's Petroleum Club in October 2012, with Clark in attendance; and a $125-a-plate dinner three months later at Calgary's Delta Bow Valley Hotel.[14]

But B.C. was not the main stage in this theatre. In industry's efforts to shape Canadian energy policy to suit its needs, billionaire Edwards had been there every step of the way. The quintessential 21st-century Canadian capitalist was made, you could say with certainty, by the oil sands.

During the years when industry backed global warming denial and voluntary efforts, Edwards was a strong federal Liberal supporter. The Canadian Association of Petroleum Producers and its allies may have lost the fight with Chrétien to prevent Kyoto ratification, but they were successful in reducing Kyoto's impacts on their bottom lines. Edwards backed Paul Martin's bid for leader and was influential in persuading the Martin government to cut industry's obligation to reduce greenhouse gas emissions from large industrial emitters by a further 45 per cent.

Then, in 2007, when industry pivoted to a national energy strategy, carbon pricing and clean growth, Edwards was in the vanguard. He was a long-standing member and director of the Canadian Council of Chief Executives, and was a member of its task force on environmental leadership that produced the "Clean growth: Building a Canadian environmental superpower" report. This document, described in Chapter 8, called for a national energy strategy, carbon pricing either through emissions trading or environmental taxation, targets and investment in technology. Most of these items would appear in Rachel Notley's climate plan.

Edwards was a participant in the 2010 Banff Dialogue that focused industry and government attention on the need for a national energy strategy. His key company, Canadian Natural Resources, was a member of the Energy Policy Institute of Canada (EPIC) and he attended meetings with federal officials as part of EPIC's efforts to persuade federal and provincial governments to move on the energy strategy.[15]

Edwards was not about to be intimidated by the New Democrats. But the Alberta Oil brand had become toxic and as a strategic thinker, he knew it. For their part, the NDP knew that the province needed to continue to rely on fossil fuels as the basis of its economy. Notley had to tread a fine line between meeting the expectations of environmentalists and those of oil executives. She would need to come up with a homegrown version of a national energy strategy to deal with climate change, that included some version of carbon pricing. Second, she would need to appeal for new pipelines, which could proceed once carbon pricing was in place. It was the same grand bargain ordered up by John Manley.

Two months after she crushed Prentice, Notley hired University of Alberta energy and environmental economist Andrew Leach to chair an advisory panel to forge a climate policy for the province. She signalled that energy would hold the upper hand in the formation of the panel. Leach was coming off a stint as the Enbridge professor of energy policy in the School of Business.

Leach may have been considered for the panel because of a column he wrote in *Maclean's* six months earlier, "Are oil sands incompatible with action on climate change?"[16] The article is a review of a study done by two British academics published in the scientific journal *Nature*. This paper concluded that, "85 per cent of (Canada's) 48 billion of barrels of bitumen reserves thus remain unburnable if the 2 degrees Celsius limit (for global temperature changes) is not to be exceeded."[17] The study received widespread positive media coverage, but Leach was skeptical. "From what I can see," he wrote, "there are significant questions with respect to field-level responses in the oil sands sector in [their] model, and I also feel they've overreached in attaching necessary conditions to oil sands production declines found in their results. Further . . . we should have serious questions about the margin of error of their predictions."

Around the same time, Leach wrote in *The Globe and Mail* that "credible climate change policy and oil sands development are not incompatible, and Alberta needs to demonstrate this to the world."[18] Notley's environment minister, Shannon Phillips, was of a similar view. "There are consequences to leaving it in the ground before a different economy is in place," she said.[19] This seemed to be the unstated mandate of the advisory panel. The door would be left open to further oil sands development.

As well as Leach's pro-development inclinations, two members of the panel were from the oil industry, one the former vice-president of sustainable development at Suncor Energy, the other the chief sustainability officer for Enbridge. What's unclear in these lofty-sounding titles is whether sustainability refers to sustaining the environment or sustaining corporate profits.

The other members were billed as the balance to oil interests, but were they? Angela Adams, a Metis woman from Fort McMurray, served as director of education for Unifor, Canada's largest private-sector union. She had been a heavy-duty equipment operator for Suncor for twenty years and was secretary-treasurer of Unifor Local 707A, the union that represented workers at Suncor and TransAlta Corporation in Fort McMurray. There had to be union representation because of the strong support unions provided for New Democrats. But Unifor 707A was dependent on the well-being of its oil and gas industry employers for its members' jobs.

The fifth member, Stephanie Cairns, brought vast experience in the fields of sustainable development, energy and climate change. She was associated for years with Sustainable Prosperity in Ottawa, the national research and policy network that promotes market-based approaches to building a lower-carbon economy. Andrew Leach as well was a member of the Sustainable Prosperity network.

No panellists were from the renewable energy sector, as Greenpeace Canada campaigner Mike Hudema pointed out.[20] Nor were any from the environmental sector. The panel was announced by the minister of

Environment, but it was not intended to balance the environment and the economy, but to move forward on the economic front. Why, otherwise, would an economist be selected to lead an environmental review?

But even before the panel began its deliberations energy company executives and clean energy advocates had already been meeting to find a way out of the impasse. One side wanted increased oil sands production guaranteed; the other wanted legislated emissions cuts. Neither was achieving its goals. They began cautious early meetings to understand the other side's position, reported *The Globe and Mail's* Jeffrey Jones.[21] A key participant was Murray Edwards, who "helped the government devise carbon and methane reduction policies . . . that reward operations that are most efficient," Jones reported. Having influenced Liberal and Progressive Conservative policy, Edwards was now influencing the NDP.

The Alberta Climate Plan was released on November 22, 2015, at the Telus World of Science Centre in Edmonton by Premier Notley and Environment Minister Phillips, but not Energy Minister Marg McCuaig-Boyd, perhaps for the optics of the event. It was an odd combination of science and wishful thinking. One journalist called it "perhaps the most ambitious package of climate change policies in Canadian history."[22] Another wrote that it "may be the most radical policy shift ever seen in this province."[23]

"I'm hopeful these policies taken overall will lead to a new collaborative conversation about Canada's energy infrastructure on its merit and a significant de-escalation in conflict worldwide about the Alberta oilsands," Notley said at the launch in Edmonton.[24] She took major steps in that direction by inviting the chief executives of four major oil sands players, Steve Williams of Suncor Energy, Brian Ferguson of Cenovus Energy, Lorraine Mitchelmore of Shell Canada, and surprising nearly everyone in Alberta business and political circles, Murray Edwards. They stood alongside leaders of First Nations and clean energy groups, as Notley announced her government's plans.

"Speaking on behalf of Canadian Natural and my colleagues from Suncor, Cenovus and Shell," Edwards said, "we appreciate the strong leadership taken today by Premier Notley [in setting] ambitious targets for the industry."[25] Steve Williams told journalist Max Fawcett that Edwards's presence was essential. "Murray was incredibly important because of his reputation for quality analysis, for detail, and for taking a long-term view. He was seen as a very important bellwether — the litmus test in this. Because he's a really tough test for most things."[26] As to what caused the dramatic shift in position, perhaps it was simply standard negotiating practice, playing the rookie New Democrats and softening up an opponent before pushing forward with what one really wants — an approach Donald Trump might recommend.

This game-changing event occurred just days ahead of a major meeting with Prime Minister Justin Trudeau and other first ministers in Ottawa in preparation for the Paris COP meeting in December.

And yet the plan wouldn't reduce emissions. The panel admitted its proposals "will not place Alberta on a trajectory consistent with global 2 degrees Celsius goals," but we're not going to do more until our peer and competitor jurisdictions act on climate change with similar policies.[27] Emissions will actually rise slightly from 267 to 270 megatonnes by 2030. If the object of the exercise was to keep global warming below two degrees Celsius, other provinces would have to do the heavy lifting, the panel argued. But the plan was designed to start the emissions curve downward in the next decades. And that would require major adjustments for Alberta consumers.

The plan consisted of four elements. Three were intended to cut emissions: putting a price on carbon, phasing out coal-fired electricity, and reducing methane gas emissions from oil and gas operations. These would be neutralized, though, by the fourth element: allowing carbon dioxide emissions from the oil sands to rise by over 40 per cent from the current 70 megatonnes a year to 100 megatonnes by 2030, which would constitute over one-third of all emissions generated in Alberta.

A carbon tax was the "backbone" of the plan. Previous Progressive Conservative government regulations placed a CAD$15-a-tonne levy on large industrial emitters (over 100,000 tonnes a year). The new plan would reach a price of $30 per tonne for both industrial and end-use emitters by 2018. The tax could continue to rise by inflation plus 2 per cent, as long as it isn't higher than taxes in Alberta's competitor jurisdictions, the panel insisted.

As earlier chapters point out, industry liked a carbon tax. Jack Mintz, who had become the President's Fellow at the School of Public Policy at the

University of Calgary and was still a director of Imperial Oil, told the panel, "when you're making investments in very expensive technologies you tend to like carbon taxes because you do get price certainty for carbon."[28]

Industry lined up behind the tax. "An efficient way to quickly target a significant and growing amount of end-use emissions would be the application of a natural gas, electricity, gasoline and diesel carbon price at the point of sale," Suncor told the panel.[29] Cenovus Energy agreed. The company "recommends the government establish an economy-wide carbon pricing system — to affect behaviour changes among all emitters in the province."[30]

So did a wide variety of organizations. "We endorse the position of economists from across the political spectrum who state that a strong and simple carbon price is the most efficient way to reduce emissions," wrote Canadians for Clean Prosperity, an organization on the right of the political spectrum with connections to the former Harper government.[31] The Pembina Institute went further than most. It recommended an economy-wide carbon tax of at least $40 a tonne, increasing $10 a year over the first ten years of the policy. According to Amin Asadollahi, Pembina's oil sands program director, a carbon tax for Alberta could generate up to CAD$9 billion in revenues and create incentives for industry and consumers to use less carbon. "If you start slow and incrementally increase it, it gives industry time to design their investments and make those decisions with the foresight of what's around the corner," he said.[32] But a carbon tax of $140 a tonne was an outlier, designed, perhaps, to frame the panel's $30 a tonne as moderate. (Panel member Stephanie Cairns is a long-time consultant to Pembina.)

Would a carbon tax actually reduce emissions? The panel thought so. "Putting a price on emissions leverages the power of markets to deploy both technologies and behavioural changes to reduce emissions over time," it wrote.[33]

The panel was guided in this conviction by the work of the Ecofiscal Commission, the pro-carbon-pricing research and advocacy group. The panel quoted the commission: "Pricing carbon emissions can help to achieve reductions at lowest cost, can contribute to global emissions reductions, and can help position Canadian firms to compete in a cleaner global economy."[34]

It sounded wonderful, but the Ecofiscal Commission presented little evidence to support these glowing tributes to the power of markets. In a 2017 study of carbon pricing policies, the commission admitted the obvious, that "the quantity of emissions reductions is not directly observable." To determine if carbon pricing actually works, researchers would have to measure the "amount emissions are reduced below the level that would have occurred in the absence of the policy." But how could one know "what emissions would have been" without the policy? As an example, the commission cited the 2008–09 economic and financial crisis that led to reductions in greenhouse gas emissions that had nothing to do with climate policy. "Emissions reductions caused by the policy can be estimated, but only with a complex energy-economy model," the commission admitted. "As in all economic models, the underlying assumptions are debatable, and thus the estimates themselves can be contentious."[35]

Not for the Alberta climate panel though. Using modelling done by Alberta Environment and Parks, but without the Ecofiscal Commission's caution, the panel concluded that, "the proposed policies would lead to emissions reductions of . . . approximately 50 megatonnes below a continuation of current policies in 2030." A significant share of these reductions would be due, not to carbon pricing, but to direct government regulation that would phase out coal-fired power and reduce methane emissions, actions the Ecofiscal Commission pooh-poohs as "command and control" regulation.[36]

The most impactful part of the plan — and the easiest to implement — may be government regulation of coal-fired power. Alberta had the highest rate of coal-fired power generation of any province. In 2013, coal generated 55 per cent of electricity in the province, and accounted for 85 per cent of the sector's carbon dioxide emissions. In 2030, the plan predicted, coal-fired electricity would be gone; the main source of power generation would be natural gas, with renewables a distant second.

Coal had already been targeted by the Harper government. Under regulations the Conservatives brought in 2012, coal-fired power plants must meet greenhouse gas emission standards or retire when they reach fifty years of operation. This was considered an adequate period for investors to earn an acceptable return on investment. Under these federal regulations, twelve of

Alberta's eighteen coal-fired plants would be retired by 2030. The other six could continue to operate; one of these, Keephills Generating Station 3 near Duffield, Alberta, until 2061. This $2 billion, 450-megawatt generating station, owned by Capital Power Corporation and TransAlta Corporation, was completed as recently as 2011.

Under the Alberta plan, decommissioned coal power plants needed to be compensated. After prolonged negotiations, the province agreed to pay hefty amounts to these companies for requiring them to shut their coal units early. Capital Power received CAD$734 million over fourteen years, TransAlta $524 million and ATCO $65.8 million. The compensation would flow from the province's carbon tax on large industrial emitters. In return, the companies were required to support nearby communities that were reliant on coal plants for their sustenance. They couldn't take the money and run. That's why it would be doled out year by year.

Shareholders liked the deal. Stock prices for all three companies jumped when the government's financial compensation package was revealed.[37] TransAlta's board of directors even awarded company CEO Dawn Farrell an extra $2.73 million in incentive compensation as part of her $7.39 million total compensation for what the board said was extraordinary leadership in negotiating the coal shutdown package.[38]

The climate plan also required the oil and gas sector to cut its methane emissions. Methane is the basic component of natural gas and is a far more potent greenhouse gas than carbon dioxide. It doesn't last as long as carbon dioxide in the atmosphere, but while it's there can have more devastating irreversible effects, such as melting sea ice, which then creates an insidious feedback loop.[39]

Methane emissions from oil and gas facilities reached 30.4 megatonnes in 2014, accounting for 11 per cent of all carbon dioxide equivalent emissions in the province. More than half came from venting, which companies do intentionally during extraction — one-third from fugitive emissions as gas leaks from equipment, and the rest from flaring, another intentional activity. The plan proposed to reduce emissions by 45 per cent from 2014 levels or 12 megatonnes by 2025. It would do this by applying new systems and technology and through a mix of voluntary practices and mandatory standards. The province released a draft proposal in April, 2018. It was criticized by environmental groups as being "considerably weaker" than regulations being considered by the federal government.[40]

Shutting down coal-fired power generation, reducing methane emissions and imposing a carbon tax, along with general energy efficiency measures, are scheduled to cut 50 megatonnes of carbon dioxide equivalent by 2030, if the climate panel's predictions can be believed. These measures will be negated by the oil sands emissions cap, allowing producers to increase their allowable carbon dioxide emissions by over 40 per cent to a maximum of 100 megatonnes a year.

Premier Notley hoped the total package would lead to a "de-escalation" of the war of words over "dirty" oil from Alberta. The province got "a wake-up call" on its environmental policies in November 2015 she said, "in the form of a kick in the teeth" from US President Barack Obama when he rejected the Keystone XL pipeline and specifically referenced the concern about importing "dirtier" oil from the oil sands.[41] Canadian Association of Petroleum Producers (CAPP) president Tim McMillan agreed that the goal of the plan was "to further enhance the reputation of our sector and improve our province's environmental credibility as we seek to expand market access nationally and internationally."[42]

Eric Reguly, the perceptive *Globe and Mail* business columnist, cautioned that we should "beware environmental announcements that the oil industry likes, and the Alberta oil industry certainly liked Alberta Premier Rachel Notley's response to her province's delinquent status on the climate file."[43] It was a fair point, especially in the energy province, but not everybody in the industry was on stage with Notley. Notably absent were oil major Imperial Oil and up-and-comer MEG Energy, for instance. As well, the oil sands sector is made up of many small and mid-sized operators who are members of the Explorers and Producers Association of Canada. They were not in attendance and did not feel the four CEOs with Notley represented them. "The group of four doesn't speak for the industry," Glen Schmidt, president and CEO of junior oil sands company Laricina Energy, objected. "The group of four speaks for themselves."[44]

According to some analysts, the 100-megatonne cap would yield one million barrels a day of additional oil sands production beyond the current 2.3 million barrels per day that produces about 70 megatonnes of carbon dioxide a year. Many of these producers worried they would be stuck with worthless assets once the cap is reached. Disagreements over CAPP's support for carbon pricing led some mid-sized producers to leave the organization and join the Explorers and Producers Association of Canada or remain unaffiliated. Others left because they felt CAPP was too focused on the oil sands, while others chose to forgo CAPP membership because, given the oil price collapse, it was too pricey for them.[45]

CHAPTER 12
NOTLEY: A NEW DEAL?

After forty-four years of Progressive Conservative rule, Alberta's business, lobbying and policy worlds "really didn't know a whole lot of people in the provincial NDP," observed Alan Ross, a well-connected partner in the Borden Ladner Gervais law firm in Calgary.[1]

It hadn't been much of a surprise when Rachel Notley appointed Brian Topp as her chief of staff. The long-time NDP strategist headed Notley's communications efforts during the election and chaired her transition team. Topp had worked for many years for federal and provincial New Democrats.

After the NDP lost the 2013 British Columbia election to the Christy Clark Liberals, Topp went into business with Conservative (Ken Boessenkool) and Liberal (Don Guy) strategists, in a firm called Kool Topp and Guy. They wouldn't lobby for clients, they explained, but would offer strategic advice on how to do it.[2] When Topp joined the Notley government two years later, the firm was renamed KTG Public Affairs, but Topp retained an undisclosed interest.[3] Two months after Topp moved to Notley's office, KTG picked up a new client — TransCanada Corporation, proponent of the Keystone XL and Energy East pipeline proposals.[4]

Despite the firm's lofty rhetoric, KTG did register to lobby the energy ministry and the premier's office, where Topp had just moved, regarding "successful construction of pipelines to export Alberta energy to market."[5] Interprovincial pipeline approval is not under provincial jurisdiction, but the company was seeking Alberta government support for the projects.

KTG no longer had a New Democrat in its operational ranks, so it hired long-time NDP insider Jamey Heath to manage the file. Heath had been Jack Layton's communications director, and earlier worked as communications director for Greenpeace. Heath had little Alberta experience, so KTG also hired former Alberta NDP candidate Ramiro Mora as an associate.[6]

The NDP's newly installed revolving door spun again as Notley brought in B.C.-based Earnscliffe Strategy Group lobbyist Marcella Munro to be her government's stakeholder engagement and communications manager in Calgary, where she would intersect with the city's oil and gas elite. Munro's public persona was as a strategist who worked on election campaigns for the B.C. NDP and Vision Vancouver, the centre-left civic party that ran city hall for a decade (2008–2018). She caused a ruckus when she turned up in Alberta because of a blog entry she posted after Vision's 2014 victory. It said that voters chose "Housing. Affordability. A greener city. Supporting our entrepreneurial culture. Saying no to projects like Kinder Morgan, to protect our environment and our quality of life."[7]

Saying no to Kinder Morgan raised red flags in the oil capital, of course. Munro claimed that her blog expressed Vision's position, not her own. It was her duty to get her guy elected, she told columnist Don Braid, but she remained 100 per cent in favour of Kinder Morgan going ahead. As a lobbyist with Earnscliffe, she said she regularly worked with B.C. oil and gas companies to get projects built. Munro could have given the example of Chevron Canada, which operated the refinery at the Kinder Morgan pipeline terminus in Burnaby on Burrard Inlet. She also lobbied for Spectra Energy on natural gas issues and BG International — taken over by Royal Dutch Shell — regarding its proposed Prince Rupert liquefied natural gas facility. Munro was later transferred from stakeholder engagement to chief of staff for the minister of Energy, where she continued to work with the fossil fuel industry. Then, in August 2017 she jumped ship to join KTG as a senior strategist. She replaced Jamey Heath, who took on a senior communications role in the Notley government. It was an exquisite case of revolving-door syndrome, one that positioned advocates for TransCanada Corporation inside and outside government in a powerful configuration.

Not that Notley needed much persuading to support TransCanada's — and other operators' — pipeline proposals. No politician who hoped to be successful in Alberta could neglect energy. Notley crafted her position in opposition, first as environment critic and then as opposition leader. "If ultimately we're going to gain support for Keystone or other pipelines like it in other jurisdictions, we need to make serious progress in terms of our environmental record," she told the Alberta legislature.[8] And in the lead-up to the 2015 election, she said, "We need to do a better job on climate change with a plan to reduce greenhouse gas emissions, and our failure to do so lies at the heart" of Obama's vow to veto any Keystone legislation emanating from the US Congress.[9]

She wouldn't lobby for Keystone XL, she explained, because it was mired in the muck of the US political system. But she would support Kinder Morgan's expansion and TransCanada's proposed Energy East pipeline to Atlantic Canada. It's a project "that we need to talk to people about and work on," Notley told the *Calgary Herald*. "Absolutely, I think there's potential there."[10]

Less than a month after the NDP won the election, Energy Minister Marg McCuaig-Boyd said her government would work with other provinces to push the Energy East pipeline forward, despite opposition from Quebec premier Phillipe Couillard, environmental groups, First Nations and residents along the proposed route. Notley had already spoken with New Brunswick premier Brian Gallant, who was a strong Energy East supporter, largely because of the terminal and shipping facility that would be built in his province. But when Donald Trump reversed directions on KXL, TransCanada withdrew Energy East.

While TransCanada snagged KTG for its Alberta lobbying, Enbridge relied on Navigator, a crisis management firm, to move the file on its bitumen pipeline. To this end, Navigator hired Sally Housser, who had been Rachel Notley's press secretary during the election campaign and who had also worked as press secretary to Jack Layton.

Along with president Ian Anderson and a roster of in-house lobbyists, Kinder Morgan retained Alberta's powerhouse lobbying firm, Global Public

Affairs. The company's list of fossil fuel company clients included oil sands giants Imperial Oil and Cenovus Energy; the Canadian Energy Pipeline Association and Alliance Pipeline; other majors like BP Canada, Chevron Resources and EnCana; as well as the Koch Companies, which had vast holdings and operations in the province but kept a low profile.

Global Public Affairs had the requisite Liberal (federal) and Conservative (Alberta and federal) connections, but didn't scramble for an entry to the NDP. Global Public Affairs' long-time vice-president for western Canada, Doug Noble, felt he could do it himself. "You've got to understand who else a [new] government may be listening to that may or may not be different than what they've listened to before," he said in an interview. "And you've got to take a look at some of those other voices and do your due diligence."[11] Besides, former Nova Scotia NDP premier Darrell Dexter was the firm's vice-chair. (The other vice-chair was Calgary-based Pierre Alvarez, former president of the Canadian Association of Petroleum Producers, while the chair was Tom Clark, former chief political correspondent for Global Television. Industry, politics, media — all bases were covered.)

In October 2017, halfway through Justin Trudeau's mandate, Global Public Affairs acquired the Wazuku Advisory Group, a British Columbia-based consulting firm with strong connections to the provincial and federal Liberal parties. It seemed to specialize in promoting the interests of companies that needed positive decisions in both Victoria and Ottawa. Wazuku functionaries had transitioned seamlessly from working for Liberal candidates during the 2015 election to lobbying the Liberal cabinet within days of its appointment.[12]

One such client was Enbridge's Northern Gateway pipeline, for which Wazuku was ultimately unsuccessful, although for reasons largely beyond the lobbying firm's control. Another client was the CAD$36 billion Pacific NorthWest LNG liquefaction and export facility near Prince Rupert. Wazuku was successful in persuading the Trudeau cabinet to issue a positive decision. However, Pacific NW, led by Malaysian-based Petronas, later decided not to proceed because of changing economic and market conditions. Then in 2017, the Liberals lost the B.C. election and Wazuku lost its preferred access to the provincial government.

Other Alberta government relations firms were quick to add New

Democrats. Impact Consulting hired Brookes Merritt, a former Alberta New Democrat communications director. Impact lobbied for Fortes Alberta and Direct Energy, providers of residential electricity and natural gas services in the province. Alberta Counsel, a consulting firm started by Wildrosers, bulked up with New Democrats, taking on former MLA Leo Piquette, campaigner Patricia Grell and former caucus outreach director Pascal Ryffel as a senior associate and later, director of government relations.[13]

Canadian Strategy Group, meanwhile, hired Moe Sihota, a former NDP cabinet minister in B.C. and Ken Georgetti, former president of the Canadian Labour Congress, as "strategic counsel." "If you take a look at what Ken and Moe are bringing to the table, it is an area that we, frankly, are lacking in," Canadian Strategy Group co-founder Hal Danchilla told *Canadian Press*. "We are looking forward to them helping educate me and everyone else in the firm, as well as from time to time giving advice to clients."[14]

Canadian Strategy Group clients included Capital Power, a major operator of coal-fired electricity generating plants in the province. This company was lobbying the government about the implementation of the climate plan, the design of future greenhouse gas regulations, wholesale electricity market design and renewable energy, among other issues. Other Canadian Strategy clients have included Athabasca Oil Corporation, a mid-sized oil sands company that purchased Statoil's oil sands holdings, and Progress Energy Resources, a leading natural gas producer owned by Petronas of Malaysia. (Global Public Affairs was also lobbying for Progress.)

However, Sihota and Georgetti appear not to have remained with the firm for long. Sihota returned to B.C. to lobby the new NDP government on behalf of Woodfibre LNG for its proposed liquefied natural gas facility on Howe Sound.

CUTTING EMISSIONS INTENSITY

Despite the sizeable minority of naysayers, the big companies were happy with the outcome of the Notley climate deal and could see a rosy future, buttressed by the persuasive power of their newly-hired lobbyists. For one thing, they were already making great strides to reduce the emissions intensity of their output. Analysts predicted the industry would be able to continue doing this even as output rises.

Steve Laut, president of Canadian Natural Resources, maintains that, "there's a very good chance that we'll never hit that cap because technology will reduce the intensity as we go forward."[15] His company cut greenhouse gas intensity per barrel 23 per cent between 2013 and 2016 by utilizing new technology at its Horizon mine, to replace one of the most wasteful aspects of oil sands extraction: burning natural gas to generate steam that is injected into underground formations to thin the heavy sticky bitumen and allow it to be pumped to the surface. Gas is a cleaner fossil fuel that should be used to heat homes, at least until it is replaced by renewable energy. Instead, it was being used to produce a dirtier fossil fuel to power trucks and cars.[16]

Other oil sands companies were moving in the same direction. In March 2016, Imperial Oil applied for regulatory approval for its Midzaghe project near Cold Lake. The project depended on new technology to cut greenhouse gas emissions by 25 per cent and double production levels. Solvents like butane and propane would be used to reduce the need for pressure and high temperatures in steam-assisted gravity drainage extraction, allowing more oil and gas to flow to the surface while using less natural gas.[17] Some companies were testing the use of radio waves instead of steam to heat sand so more oil could flow to the surface.

The industry's move to cut emissions and boost clean technology, while increasing oil sands production was supported by the federal and Alberta governments. In November 2017, the two governments announced CAD$28.8 million for so-called clean technology projects. Most of this spending would be for oil sands production, including $10 million for energy efficient ways to explore for oil and $8.2 million to develop a new process to cool steam created during production.[18]

A 2017 CIBC research report concluded that, "under the current technology Alberta's emission cap . . . would be reached in the second half of the next decade, but our analysis suggests there is the potential for this to be pushed out much further in time."[19] A study by the Canadian Energy Research Institute agreed. "With the various technologies that are available for commercialization within the next five to seven years, industry is able to grow production and incremental barrels of bitumen without reaching the emissions cap," said Dinara Millington, the institute's vice-president of research.[20]

About eight months after standing on-stage with Notley and endorsing the 100-megatonne cap, Suncor Energy's Steve Williams pledged to cut the emissions intensity of the company's oil and petroleum products by 30 per cent by 2030. "Climate change is happening. Doing nothing is not an option," he said as part of the company's annual sustainability report. The goal he set to reduce the "emissions intensity" of operations referred not to an absolute decrease, but to the amount of emissions per unit of output. Suncor's total greenhouse gas output would still rise because it produced more barrels of oil.[21]

Suncor's strategy for growth is important because it is the descendant of the very first oil sands surface mining operation, the Great Canadian Oil Sands plant that opened in September 1967. It celebrated its fiftieth anniversary in September 2017, so longevity is written into its DNA.

In 1979, Sun Oil created Suncor Energy by merging all of its Canadian holdings. Then, in 2009, Suncor merged with Petro-Canada to create the country's largest energy company. Finally, in 2016 Suncor completed a takeover of Canadian Oil Sands for CAD$4.2 billion, whose main asset was a 37 per cent stake in rival oil sands producer, Syncrude, the largest producer of synthetic crude oil in Canada. Suncor already had a 12 per cent stake in Syncrude and after buying Murphy Oil's 5 per cent share for $937 million, Suncor ended up as the majority owner of Syncrude.

Much of Suncor's expansion was driven by mergers and acquisitions, so there was no net increase in greenhouse gas emissions. But emissions from new capacity would be an issue. As of 2018, Suncor was the main partner in the CAD$15.1 billion-plus Fort Hills mine at the northern end of the Athabasca oil sands region, ninety kilometres north of Fort McMurray. When fully operational, the mine will produce 180,000 barrels of bitumen a day. Other companies, such as Imperial Oil and Canadian Natural Resources, are also adding new capacity.

To deal with emissions from the Athabasca Oil Sands Project, Shell Canada developed the CAD$1.35 billion Quest carbon capture and storage facility near Fort Saskatchewan, Alberta, largely with federal and provincial funds, to bury 30 million tonnes of compressed carbon dioxide two kilometres underground. (This approach is prohibitively expensive and no use is made of the carbon dioxide.)

Industry also spent money to find profitable uses for carbon emissions.

Eight oil sands companies including Shell, Imperial, Canadian Natural Resources and Suncor, invested a total of CAD$10 million in the Canada's Oil Sands Innovation Alliance's Carbon Xprize in 2015. Another $10 million came from NRG Energy, an American electricity generation company. Teams from around the world signed up for the four-and-a-half-year competition.[22] The winning team would be the one that converts the most carbon dioxide into products with the highest net value and the smallest environmental impacts. Federal and Alberta governments kicked in up to $10 million each to develop the Alberta Carbon Conversion Technology Centre in Calgary, where carbon conversion technologies could be tested on a commercial scale. The contest finalists would be the first tenants of the facility.[23]

New technologies and government support could prevent the oil sands sector from breaching the 100-megatonne ceiling. But as more money and effort goes into cutting emissions intensity per barrel of heavy oil and into finding profitable uses for carbon, 100 megatonnes per year would become entrenched in Canada's carbon budget, extending far beyond 2030 given the vastness of the bitumen resource. If the bitumen continues to be available, it will be used, and a low-carbon future that much less likely to be achieved.

NOTLEY AND ROYALTIES; DOES SHE OR DOESN'T SHE?

Notley promised action on climate change to break the deadlock between environmental critics and industry advocates; she also promised a review of the province's royalty structure for oil and gas. Notley charged that Albertans were not receiving full and fair value for their resource because the province had one of the lowest royalty structures in the world (thanks to the scheme put in place by the Ralph Klein government). The energy industry was filled with fear and trepidation over the vision of New Democrats setting royalty rates — just another cash grab, they reckoned.

The NDP leader had not even taken office before energy executives were crafting positions and lining up lobbying teams. Most lobbying would take place through the Canadian Association of Petroleum Producers and its stable of thirty-five lobbyists led by association president Tim McMillan. Global Public Affairs was registered to lobby for BP and gas pipeline operator, the Williams Companies. Hill+Knowlton Strategies was registered for

Statoil. Other companies used their in-house lobbyists to get their message across.[24]

They geared up for battle, but no battle occurred. Six months and sixty-five stakeholder meetings later, the royalty review panel Notley had appointed concluded that Albertans were receiving their fair share, oil sands royalties wouldn't change, existing rates would apply to conventional oil and gas wells already operating, and only new wells would have a revised rate structure.

The industry was definitely pleased. As the CBC's Tracy Johnson saw it, "it was the best possible outcome for the energy sector."[25] The government's status quo responses to royalty questions "are signals that [it] is serious about encouraging investment in Alberta at this difficult time," Tim McMillan said, as if encouraging investment in the oil sands and not receiving full and fair value should be the goal of royalty policy.[26]

The panel's conclusions may not have been surprising given its makeup. It was led by the head of provincially owned ATB Financial, which lends money to energy companies. Its members included the chief economist at ARC Financial Corporation, an investment firm with stakes in energy companies, and a former Alberta deputy minister of Treasury Board and Finance, who oversaw previous royalty giveaways to the industry.[27]

The review was undertaken at a time of low oil prices, so it made sense to retain low rates. But these rates will continue regardless of the price of oil. Under the previous Progressive Conservative regime, energy companies were seriously undercharged for their use of the resource, resulting in billions of dollars of lost revenues for the provincial treasury, a charge made by former senior advisor on royalty policy for Alberta Energy, Jim Roy. "Nobody talks much about the government gifting the petroleum industry $13 billion," he said.[28]

As Ricardo Alcuna, executive director of the progressive Edmonton-based Parkland Institute, explained, "it is a fundamental principle of royalties that the bulk of economic rent resulting from high prices should go to the resource owners, not the industry. The reason for that is the high prices and increased rent are directly a result of the value of the resource itself, not anything the companies extracting it have done."[29]

The most stinging rebuke came from the NDP's long-time ally, the

Alberta Federation of Labour. "What happened can best be described as the government being captured by industry," federation president Gil McGowan remarked. Fair enough, there should be no raises in royalties when prices are low. "But many experts have agreed our system was broken at the high end," he said. "Unfortunately, what was announced . . . does nothing to address the problem."[30]

THE OIL SANDS ARE CANADIANIZED

The makeup of the Alberta oil sands industry was undergoing dramatic changes when Notley took charge. Most foreign-based companies were exiting the field, or about to. Canadian-based companies also changed, some growing into much larger entities by buying up foreign firms' holdings.

When they agreed to a carbon tax and a cap on emissions, Suncor Energy and the rest seemed to be signing their death warrant. How could they survive? Yet they ended up larger and stronger than ever. Most of their shareholders resided outside the country, but they were headquartered in Calgary. As for the global companies, their oil sands operations, though large in dollar terms, were sideshows, add-ons to their global ambitions.

Then, in December 2016, Statoil sold its entire oil sands holdings to Athabasca Oil Corporation, a smaller Canadian-based producer, for up to CAD$832 million, depending on the price of oil.[31] In February 2017, ExxonMobil announced the largest write-down of reserves in the company's history, removing the entire US$16 billion 3.5 billion-barrel capacity recently completed Kearl oil sands mine from its books.[32]

But the main event was Canadian Natural Resources' purchase of Royal Dutch Shell's oil sands assets. It caught everyone's attention. For maximum impact, the deal was announced in March, 2017 at the CERAWeek conference in Houston, which was attended by everyone who's anyone in the oil industry. *The Globe and Mail*'s Calgary energy reporter, Jeffrey Jones, suggested Canadian Natural Resources' chairman, Murray Edwards, "made the approach, sensing an opening as Shell looked to sell US$30 billion of assets worldwide. His timing was perfect."[33] Not so much for Justin Trudeau, though, the first sitting Canadian prime minister to address the conference. Shell's pull out from the oil sands had to hurt his message that investing in Canadian energy was a no-brainer.

Canadian Natural Resources was paying US$8.5 billion, in cash ($5.4 billion) and shares (98 million, or 9 per cent, worth $3.1 billion), for Shell's 60 per cent share of the giant Athabasca Oil Sands Project, with its capacity of 255,000 barrels of oil equivalent per day. The purchase included a mine north of Fort McMurray and the Shell-operated Scotford bitumen upgrader and Quest carbon capture and storage facility northeast of Edmonton, although Shell would continue to operate these.

In a second deal, Canadian Natural Resources and Shell paid US$1.25 billion each for Marathon Oil Corporation's 20 per cent stake in the Athabasca Oil Sands Project, giving Canadian Natural Resources 70 per cent, Shell 10 per cent (plus 9 per cent of Canadian Natural Resources' share), and Chevron Canada, the remaining partner, with the other 20 per cent. Marathon lost money on the deal — it paid US$6.9 billion for Western Oil Sands' 20 per cent stake in 2007 — but wanted to cash in so it could invest $1.1 billion in the Texas Permian Basin shale play.[34]

The two deals cemented Canadian Natural Resources as one of a handful of companies — the others being Suncor Energy, Imperial Oil and Cenovus Energy — that came to dominate the mining and processing end of the bitumen business. Except for Imperial, these are the companies that stood behind Alberta's climate plan. The two deals added about 50 per cent to Canadian Natural Resources' reserves and boosted overall production to one million barrels per day (BPD), making Canadian Natural Resources the country's second-largest energy company. The deal was a particularly good one for Canadian Natural, since it was buying new capacity for 40 per cent less than the development costs at its expanding Horizon mine, across the Athabasca River from the Athabasca Oil Sands Project. And it enjoyed the benefit of immediate cash flow.

Three weeks after the Canadian Natural Resources-Shell blockbuster, an even larger acquisition occurred when Cenovus Energy paid ConocoPhillips CAD$17.7 billion in cash and shares for the Houston-based company's 50 per cent share of their joint venture, a steam-driven operation south of Fort McMurray, plus gas and oil assets in the Deep Basin region of Alberta and B.C. The acquisition would allow Cenovus to double its total output to 558,000 barrels of oil equivalent a day from its now wholly owned operations to become the third largest oil sands producer. ConocoPhillips ended up as the largest Cenovus shareholder, with a 20 per cent holding.[35]

Investors didn't like this deal because to pay for it the company issued shares—diluting their holdings — and took on a large debt load — creating uncertainty about future prospects.[36] The view on Bay Street was that the company paid too much. Then, several months later, sixty-year-old CEO Brian Ferguson retired from the company, sending shares into a further tailspin. But by the end of the year, sales and profits were up and the company planned to sell CAD$5 billion worth of assets to pay down debt.[37]

SHELL CHANGES DIRECTION

The transfer of important assets from a foreign-owned to a Canadian-owned company would, once upon a time, have excited Canadian nationalists concerned about foreign domination of the oil patch.

Exxon tended to be the nationalists' whipping boy, but the tale of Shell in Canada says at least as much as that of Exxon about oil geopolitics. As big as the oil sands are in the Canadian economy and Canadian politics, it barely rates as a pawn on Shell's global chessboard.

The sale to Edwards reflected Shell's decision that the future lay in natural gas. This allowed it to promote the idea it was moving toward a low-carbon future, although natural gas will never get the company there. It made a major move in that direction with the US$52 billion takeover of BG Group, one of the world's largest natural gas companies.

To pay for this massive acquisition, Shell committed to off-loading US$30 billion worth of assets. The oil sands sale would contribute nearly 30 per cent of this amount. The exit signalled the end of an era: Shell's had been the third major project to move forward in the oil sands in the 1990s, after Great Canadian Oil Sands in the 1960s and Syncrude in the 1970s, and before the stampede of foreign investment hit the province after the turn of the millennium.

Although it was now out of the oil sands, Shell was still a big Canadian player, with substantial natural gas assets. BG Group's major Canadian project was its proposed CAD$16 billion liquefied natural gas export terminal near Prince Rupert, B.C. But Shell was already a 50 per cent partner in the rival $40 billion liquefied natural gas export terminal proposal near Kitimat, with PetroChina Co., Korea Gas Corporation and Mitsubishi

Corporation.[38] After the oil sands sale, Shell shuttered the BG plant and was still "very much" considering LNG Canada.[39]

When Shell developed the Athabasca Oil Sands Project, it was as a Canadian company called Shell Canada. When it sold the assets in 2017, it was as international behemoth Royal Dutch Shell. Despite the name Shell Canada Ltd., the company existed as a legal entity on paper only, a wholly owned subsidiary of Royal Dutch Shell plc. headquartered in The Hague. The switch was made in 2007, when Royal Dutch Shell paid CAD$8.7 billion for all the shares of Shell Canada it didn't already own. Shell Canada went from being a public company with an independent board of directors and its own CEO to a small division of the global giant. Operations headquarters were moved to Houston, Texas, while major decisions were made in The Hague and London. Where before Shell Canada had its own CEO, it now had a country chair, Lorraine Mitchelmore, whose actual operational job was continental, not Canadian: executive vice-president of heavy oil for Shell's upstream Americas division. When Mitchelmore retired in 2015, Canada was crunched further in the organizational chart in Houston.

The parent company may have executed the buyout to gain control over the Athabasca Oil Sands Project. In the glory days of the mid-2000s, when oil prices were skyrocketing, Shell had ambitious plans to expand bitumen production from 155,000 to 770,000 barrels per day, and likely didn't want Canadian interests on the board presenting roadblocks. "Instead of building expensive upgraders in Alberta, [Shell would have] the flexibility to move the bitumen to the Gulf Coast," one Calgary analyst speculated at the time, although he couldn't have been aware of the pipeline hurdles that would be erected in its path.[40] If Shell Canada had had its own independent board, it might have wanted to build such upgraders for the full-time jobs and economic benefits they would bring to the province and country.

CHAPTER 13
WELCOME TO THE CLEAN GROWTH CENTURY

When Rachel Notley climbed onto the Telus World of Science stage on November 22, 2015 to announce her climate plan, she had some critical information in her back pocket.

That Notley could proceed with confidence had to do with a dinner that occurred in Ottawa just days after Justin Trudeau's election victory in October. As *Vancouver Sun* reporter Peter O'Neil revealed, the party comprised three Alberta NDP insiders and an equal number of key figures in the new Trudeau government.[1]

Representing Alberta were Brian Topp, Notley's chief of staff; Richard Dicerni, head of the Alberta public service; and Andrew Leach, who was then putting the final touches on the climate plan. Sitting on the other side of the table were Trudeau's top political operatives, Gerry Butts and Katie Telford, along with Privy Council clerk Janice Charette.

After they discussed Alberta's miserable reputation on the international climate change front, it became clear that Trudeau would back Notley's carbon tax and "take it national" in short order. And the national carbon tax would give Trudeau licence to approve pipelines that would expand Alberta's oil sands production — a grand bargain in the making. Knowing that Trudeau had her back, Notley could proceed with her plan. The stars were aligning nationally and provincially: Trudeau and Notley could count on each other's support during the carbon pricing and pipeline approval wars.

It didn't take long for events to unfold. Two weeks later, John Manley, head of the Business Council of Canada, gave Trudeau some advice on how to get serious about climate change.[2] Writing in the online magazine *ipolitics*, Manley reminded Trudeau that the chief executives had been on record since 2007 about the need to put a price on carbon. He then made two points: Trudeau had to demonstrate a commitment to "responsible" climate action and he needed to step up efforts to support the export of energy products. More pipelines please. And from the actions Manley said must be undertaken — don't damage the competitiveness of Canadian companies, phase in carbon pricing gradually, use revenues raised primarily to cut corporate and personal income taxes — it's clear the responsibility was to the financial well-being of Canadian companies and not to the future of the planet.

And that's what Trudeau did over the next year, demonstrating a commitment to "responsible" climate action without damaging the corporate bottom line, an agenda also followed by Notley. On the export side of the equation, Trudeau approved two diluted bitumen pipelines plus a liquefied natural gas plant on the British Columbia coast. But he rejected Enbridge's Northern Gateway pipeline which by this time was clearly dead to everyone, probably including Enbridge. He also imposed a ban on oil tanker traffic from Quadra Island to north of Prince Rupert, a decision that was necessary to have any hope of moving forward on the national energy strategy. The day before Trudeau approved the pipelines in November 2016, Manley restated his two points as a "grand bargain": acquiescing to a price on carbon on one side, building pipelines on the other.[3] Manley reminded all and sundry that the Business Council of Canada had signed on to carbon pricing, so long as it meant getting resources out of the ground and to their customers. "I would be very surprised if there were no pipelines emerging from this," Manley said at a business forum in Lake Louise, Alberta sponsored by law firm Bennett Jones. "Acquiescence to a price on carbon really is looked at as one side of a grand bargain that would see pipelines built in return."[4] And the day after Trudeau approved the pipelines, Manley applauded, claiming that approving the pipelines "balances strong environ-mental protection with responsible energy development."[5]

There was more to this seeming convergence between CEOs and Trudeau. The Liberal government's blueprint for a low-carbon future, the "Pan-Canadian Framework on Clean Growth and Climate Change," was eerily similar to the declaration of Manley's group nine years earlier, "Clean Growth: Building a Canadian Environmental Superpower." Aside from a focus on clean growth — a declaration that growth will continue whatever "clean" comes to mean — the parallels in the documents are remarkable. The chief executives demanded a "coherent national plan of action"; Trudeau delivers a pan-Canadian framework, not quite a national plan, but on the way. The chief executives asked for investment in new technologies; Trudeau delivers investment in clean technologies. The chief executives demanded price signals such as emissions trading or carbon tax; Trudeau delivers carbon pricing through either emissions trading or carbon tax.[6]

The long-awaited national energy strategy was rolling into the station. Trudeau had already adopted the business framing of a transition to a low-carbon economy as an opportunity. In preparation for meetings in Vancouver with provincial and territorial leaders, he said the talks "will focus on effective ways to . . . capitalize on the opportunities presented by a low-carbon economy," not on what we must do to keep global warming within two degrees Celsius.[7]

In Paris, at the make-or-break climate change meetings, the talk was all about two degrees Celsius and even 1.5 degrees Celsius, a vastly more ambi-tious target promoted by Trudeau's Minister of Environment Catherine McKenna. It was a target Canada had no intention of meeting, as became apparent over the next year. Canada's goal was to cut greenhouse gas emis-sions — its intended nationally determined contribution — 30 per cent below 2005 levels by 2030, reducing emission from 742 megatonnes to 517 megatonnes, admittedly a daunting task (projected as of December 2016). Yet while McKenna was setting praiseworthy temperature and emis-sion-reduction targets and Trudeau was telling the assembled dignitaries that "Canada is back," McKenna was designing Canada's escape hatch as well, as chair of the Article 6 committee that authorized emission markets.

SUSTAINABLE PROSPERITY SMARTENS UP

Trudeau had barely returned from his European jaunt when he received a letter from a new organization calling itself Smart Prosperity. This "diverse group of leaders from business, labour, aboriginal and environment perspectives" warned him that:

> As the world grapples with climate change and other growing
> environmental problems, the global economy is transforming.
> Before long, the strongest economies will be those that have
> been rebuilt to run on clean energy, conserve resources, reduce
> waste, provide vibrant and healthy communities for people to
> live in, and preserve nature. The world around us is changing,
> and Canada will either shape change or have it buffet us.
> We applaud the steps you have already taken to change the
> conversation about these choices, and to establish Canada as
> a voice for positive action around the world. The steps your
> government takes — whether on climate policy, investing
> in infrastructure and skills, or supporting science and clean
> innovation — all will set a tone and a pace for the country.[8]

Smart Prosperity volunteered its services to help Trudeau achieve the lofty ambitions laid out in the letter. Smart Prosperity was a virtual organization, based largely on Bay Street, with the University of Ottawa's Sustainable Prosperity Research and Policy Network providing its research and secretarial services. With the formalization of the relationship in 2016, the academic wing rechristened itself as the Smart Prosperity Institute. The result was an organization that blurs the lines between government, academia and business, similar to the entities created by Bruce Carson. Much of the Smart Prosperity Institute's funding came from government; other funding came from charitable foundations like the Jarislowsky and Atkinson foundations and the private sector, although these latter sources are not identified.

This was an unusual activity for an academic institution. It seemed to be a strategy to shift public and political attention away from the need to drastically cut greenhouse gas emissions and on to some nebulous clean energy future. Federal government bureaucrats, university academics, business executives and representatives from non-governmental organizations were being brought together under one umbrella emblazoned with the word "prosperity."

Trudeau must have liked what he read, because he was guest speaker three months later in March 2016 when Smart Prosperity officially launched itself in Vancouver, giving the group semi-official recognition. "The government of Canada stands with you," he pledged.[9] Just how closely would be revealed in the months to come. It was an impressive achievement for the fledgling organization, enabled perhaps by its close connections to Environment Canada and the Trudeau government.[10] There was Velma McColl, the former Liberal adviser and her Shell Canada client. (Shell Canada was a Smart Prosperity funder and former Shell Canada country chair Lorraine Mitchelmore a Smart Prosperity co-chair. McColl was also on the executive committee of the Business Council of Canada.) In 2018 other Smart Prosperity leaders included:

- Dominic Barton, managing director of global consulting giant McKinsey & Company, with some of the world's largest corporations as clients. Barton had just been appointed to head the Liberal government's economic advisory council;
- David Miller, former Toronto mayor, and president and CEO of World Wildlife Fund Canada. Gerald Butts, his predecessor at WWF, is Trudeau's principal political adviser;
- Ed Whittingham, executive director of the Pembina Institute, succeeded Marlo Raynolds, who was appointed by Trudeau to be chief of staff to Environment and Climate Change Minister Catherine McKenna.

To accompany the launch, Smart Prosperity issued a report designed to "harness new thinking" that would map out and accelerate Canada's transition to a stronger, cleaner economy in the next decade. In this new thinking, no mention is made of Big Oil, the primary cause of global warming. In Smart Prosperity's world, there are no pipelines, no oil sands, no fracking, no First Nations, no poor, no workers. And there is no role for direct government participation, let alone leadership, in ending reliance on fossil fuels. Instead, government can fund education and research and development. It can also work with private-sector partners and markets to promote economically promising areas of innovation, and craft policies to accelerate improvements and help private sector firms overcome cost barriers. Government can put a price on pollution to create market rewards for better choices. It can even participate in public-private partnerships to improve infrastructure.[11] "Accelerating clean innovation" is a major focus.

THE NATIONAL ENERGY STRATEGY COMES IN FROM THE COLD

The day after Trudeau expressed solidarity with Smart Prosperity, he met with provincial and territorial leaders to launch the long-awaited negotiations on the national energy strategy, now known as the Pan-Canadian Framework on Clean Growth and Climate Change.[12]

The premiers had worked on the strategy since Alison Redford jump-started the process in 2012. Redford's advocacy in turn stemmed from Canada West Foundation president Roger Gibbins's 2007 paper and that in turn emanated from Enbridge CEO Pat Daniel, who in 2006 mused about how his Northern Gateway pipeline needed to get buy-in from other provinces and First Nations. Redford was gone now, but Rachel Notley was in, and the premiers soldiered on, finally releasing the strategy at a meeting in St. John's in July 2015. By then it had been transformed into something akin to John Manley's "grand bargain": fast-track the construction of energy pipelines and reduce greenhouse gas emissions. A promise that all provinces would adopt "absolute" cuts to greenhouse gas emissions was stripped from an earlier version of the plan, leaving little more than a series of vague aspirations for a "lower carbon economy."[13]

Trudeau had started using the term "pan-Canadian framework" to refer to the national energy strategy eight months before the 2015 federal election. It was an important framing device for him. He had to overcome his father's legacy regarding Ottawa-centric policy that was seen as hostile to Alberta's oil and gas wealth.[14]

The framework document was released in December 2016 after nine months of negotiations following the federal-provincial meetings in Vancouver. Its ostensible purpose was to lay out a strategy for getting to Canada's greenhouse gas emissions target — a 30 per cent reduction from

2005 levels by 2030, cutting 219 megatonnes a year by then. This was the Harper target. It was pooh-poohed by the Liberals in opposition as too little. Once in power, the Liberals said that 30 per cent would be a floor they would easily exceed.

In the Pan-Canadian Framework, though, it's a ceiling they won't reach. The commitment should have been front and centre to the federal-provincial effort. But it wasn't. The pathway to meeting Canada's 2030 target is presented on page 45 of the 47-page Pan-Canadian Framework document. No references or sources of information are provided. Many ways of reducing emissions are on the shopping list along with carbon pricing: purchasing carbon credits from California, cutting methane emissions, phasing out coal-fired power generation, bringing in a clean fuel standard, and implementing stricter federal building codes.

Measures already announced by federal and provincial governments and those included in the framework document, as nebulous as they are, would result in a shortfall of 44 megatonnes (Mt) that must come from somewhere else — called "additional measures" — if the numbers can be trusted. These could be investments in public transit, green infrastructure and clean technology. Cuts won't be coming from the oil and gas sector, though, the source of one-quarter of all emissions. True, Alberta did legislate a cap of 100 Mt on oil sands production, but since emissions were at 70 Mt in 2013, and even less in 2005, this is a substantial increase. The framework adds little else to control this growing source of emissions.

Media attention focused almost obsessively on the price Trudeau was putting on carbon — CAD$50 a tonne by 2022. Simon Fraser University energy economist Mark Jaccard, an expert in carbon pricing, suggested the price needs to be $160 a tonne by 2030 to get the country to its target — a political impossibility, at least in 2018.[15] Canada may have to purchase international carbon credits to meet its target.[16]

And that may be the true bottom line. Ottawa would be working at the UN to establish the rules for international trading that Environment Minister McKenna helped bring into the Paris Agreement. But if Canada ends up relying too heavily on foreign credits and offsets, then not much will really change except for an increase in funds flowing to California and developing countries where Canada can buy offsets. And the closer

Canada gets to more pipelines for oil sands exports, the more likely this is to happen.[17]

Parties are required to file reports to the United Nations Framework Convention on Climate Change every two years. Canada's 2017 report bragged about the strong progress the country was making toward meeting its Paris goal. The list of actions under way was long and impressive, but the bottom line was that the national effort would fall far short of the target, which was 517 Mt of greenhouse gas emissions by 2030.[18] Instead, the country was heading toward 583 Mt, creating a gap of 66 Mt.

As Canada was submitting its report, the Organization for Economic Cooperation and Development weighed in with its report card on Canada's environmental progress. The key finding was that Canada could not meet its goal unless it made significant cuts to emissions from the oil sands. "Without a drastic decrease in the emissions intensity of the oilsands industry, the projected increase in oil production may seriously risk the achievement of Canada's climate mitigation targets," the report noted.[19] "Canada is the third largest emitter [in the OECD] in per capita terms — and emissions show no clear sign of falling."

Canada's own greenhouse gas inventory, filed several months later, showed just how far off the mark the country was from meeting its target. Canada had cut its emissions by 1.4 per cent between 2015 and 2016, with most of the decrease resulting from switching from coal to gas and other sources of fuel. At the same time, emissions from oil and gas production rose by nearly a half since 2005.[20]

The oil sands problem must have been a hot topic of conversation during the October 2015 dinner attended by Notley's and Trudeau's emissaries. Notley couldn't get carbon pricing from Big Oil unless she granted a significant increase in oil sands production. Trudeau couldn't get pipelines unless he could bring in carbon pricing. It was a perfect circle. "We all know that we need to get beyond fossil fuels," Trudeau conceded in Vancouver in 2016. "We are simply not there yet. And in the meantime, in the transition, not only do we need jobs and growth in the economy, we need to be figuring out how to extract and develop our natural resources, including fossil fuels, in smarter, cleaner, more responsible ways."[21] And if we need to develop our natural resources, we'll need pipelines. It's as simple as that.

Just don't talk about pipelines when you're discussing climate change. Neither the "Pan-Canadian Framework on Clean Growth and Climate Change" nor its accompanying report, "Federal Actions for a Clean Growth Economy" mentions the word pipelines.

PIPELINE POLITICS

"Pipelines may be straight but the stories behind them have many twists and turns," Trudeau's first Natural Resources Minister, Jim Carr acknowledged.[22] He should know. He'd been helping to write the oil sands pipeline story since 2009 when he co-sponsored the Winnipeg Consensus, the meeting of think tanks that kicked off work on the national energy strategy. In his job as minister, his department was the target of persistent pipeline company lobbying as his government made up its mind about the projects it would approve. In the year leading up to that momentous decision in November 2016 when Trudeau approved two pipelines and a liquefied natural gas plant, the big three pipeline companies met with officials from Carr's department nearly sixty times, or just over once a week.

The effort seemed to have paid off. Enbridge lost Northern Gateway, but it got its Line 3 replacement; Kinder Morgan got its Trans Mountain expansion; TransCanada got nothing, at least not immediately. All three companies still had skin in the game. If their projects all proceeded, 1.2 million barrels per day of new capacity would be added to an existing system that transports 2.4 million barrels per day.

But by 2018, they had all run into difficulties. Big oil sands producers were cutting back on their investments in new production and instead were putting their cash into dividend payments and share repurchases. "We don't see major investments in the Canadian oil sands until we see an improvement in the competitive position of the industry," Suncor Energy chief executive Steve Williams told industry analysts. We need "to see shovels in the ground and pipelines being built before contemplating new investments," he added.[23]

Enbridge was already building the Canadian section of Line 3, but faced obstacles for the American section, which traverses the state of Minnesota before ending up at Superior, Wisconsin. The company's plan was to not follow the existing right-of-way but to develop an alternate route and leave the existing line in the ground. The reason was mostly financial. Enbridge estimated the cost to remove Line 3 to be US$1.3 billion compared with $85 million to simply decommission and abandon the line. In April 2018, a Minnesota judge ruled that Line 3 could proceed, but only along the existing right of way and only on condition the company remove and replace the aging line. The recommendation from the judge was not binding but would provide input into a final decision by the state's Public Utilities Commission.[24]

This may not have been overly worrying for Enbridge, given that in 2017 alone it spent more than US$5 million to influence the commission, if one includes fees paid to lobbyists and lawyers to argue Enbridge's case before the commission.[25] Finally, at the end of June 2018, the commission approved the project over the company's preferred route. It was high-fives all around as Enbridge shares rose 6 per cent in mid-day trading on the Toronto Stock Exchange and ended the day up 3.7 per cent.

TransCanada faced similar hurdles with the American section of its Keystone XL line to Steele City, Nebraska. Even with Donald Trump's blessing, the line was confronted by vigorous grassroots opposition in Nebraska and Montana. The company, however, received a favourable decision from the Nebraska Public Service Commission that the project was in the state's interest. This was the last major approval needed from state regulatory bodies. But, like Minnesota's Public Utilities Commission, the Nebraska agency did not approve the company's preferred route. Instead it green-lighted a so-called "mainline alternate route," which was slightly longer.

Both Rachel Notley and Jim Carr applauded the decision.[26] Farmers and landowners along the new route were not as enamoured. They filed an appeal in the Nebraska courts at the end of December 2017. TransCanada began negotiations with the landowners. Meanwhile the company was lining up shippers to take space in the pipeline, booking 60 per cent of the pipeline's 830,000 barrels per day capacity.

But it was the Trans Mountain pipeline that attracted the most heated opposition. That the fix was in for this pipeline to proceed was revealed via an investigation by Mike De Souza of the *National Observer*. Through interviews with government insiders and documents obtained through freedom of information requests, De Souza unearthed the story of how a senior official in Carr's department told the team of government bureaucrats working on the file "to give cabinet a legally sound basis to say 'yes' to Trans Mountain."[27]

These instructions came a month before Trudeau did say yes — during a time when the government was supposed to be consulting in good faith with First Nations, who were led to believe the government had not yet come to a final decision. The marching orders were delivered by Erin O'Gorman, Associate Deputy Minister of Natural Resources, who led the Major Projects Management Office, an interdepartmental office set up by the former Harper government to speed up federal review of major projects.

Kinder Morgan had lobbied O'Gorman four times over the year, with one contact occurring two weeks before the meeting in which she conveyed government's need for a yes. With its in-house stable of a dozen registered lobbyists, Kinder Morgan lobbied O'Gorman's boss, deputy minister Bob Hamilton, six times and Carr's chief of staff, Janet Annesley, twice. The company didn't lobby Trudeau directly but was registered as having met four times with Trudeau's principal secretary, Gerry Butts.

There should have been nothing alarming about the O'Gorman directive. Given the grand bargain formulated by the Business Council of Canada and adopted by Trudeau and Notley at the Ottawa restaurant in 2015, getting to yes on Trans Mountain was an essential element in the deal. "We stand by our decision [on Kinder Morgan]," Environment and Climate Change Minister Catherine McKenna said at the 2018 Globe Forum in Vancouver. "We would not have been able to get a national climate plan if we did not have the key support of provinces like Alberta." And that support depended on Trans Mountain.[28] Trudeau pledged to get a pipeline to tidewater built.

Getting to yes within government was still a long way from getting to yes on the pan-Canadian playing field. The most troubling example of how O'Gorman's order short-circuited the Aboriginal consultation process occurred on November 28, 2016 when Chief Maureen Thomas of the Tsleil-Waututh First Nation on B.C.'s Burrard Inlet — ground zero for the pipeline — submitted four reports to Carr. Comprising 164 pages, the reports raised numerous concerns about economic and environmental impacts. Public servants took less than a day to review and reject all four submissions, De Souza reveals. Later that afternoon Trudeau, Carr and other cabinet ministers announced their approval of the pipeline expansion.

The Tsleil-Waututh people joined with other First Nations, whose traditional territories would be crossed by the pipeline, and environmentalists
who feared the despoliation of the Salish Sea in a well organized and well
funded campaign of resistance. Thousands marched, hundreds were arrest-
ed, including Green MP Elizabeth May and NDP MP Kennedy Stewart, who
was preparing to throw his hat in the ring for the Vancouver mayoralty race,
and scores of protesters set up a permanent camp near the entrance to the
Kinder Morgan worksite. They vowed to continue their protest for as long
as it took to defeat the project.

A second and perhaps more impactful source of resistance came from
B.C.'s New Democratic government. The preceding Liberal government led
by Christy Clark had given its support for the project after Kinder Morgan
met five conditions, including a revenue-sharing agreement worth up to
CAD$1 billion. But after the May 2017 provincial election, the B.C. New
Democrats and the Green Party of B.C. entered a power-sharing arrangement and agreed "to immediately employ every tool available" to stop the
project. In February 2018 B.C. Premier John Horgan said he would take a
reference case to the courts on barring additional diluted bitumen ship-
ments through B.C. until further scientific studies were completed on how
to clean up a spill.

Facing uncertainty, especially over the B.C. court challenge, Kinder
Morgan announced it would walk away from the project at the end of May
2018 if it didn't see a clear way to completion. As the deadline loomed, the
Trudeau government announced it had reached a deal with Kinder Morgan
to purchase the project for CAD$4.5 billion.

The purpose of the action was to keep construction moving forward with
the intention to sell the project back to the private sector as soon as possible. Should that ever happen, it would be interesting to see what price
Ottawa fetches, given that Kinder Morgan valued the aging system at $550
million in 2007, as Andrew Nikiforuk reported in The Tyee.[29]

Still, for a moment it may have seemed that Justin Trudeau was taking a
similar course as his father when he created Petro-Canada forty-two years
earlier. It's clear, though, that there are few similarities. One was Keynesian
in intent, the other neoliberal. Pierre created Petro-Canada to benefit

Canadian consumers and to promote the Canadianization of the oil industry. Justin purchased Trans Mountain to bolster the prospects of the Alberta oil sands industry.

THE CLEAN GROWTH CENTURY ARRIVES

The "Federal Actions for a Clean Growth Economy" report accompanying the Pan-Canadian Framework document laid out the federal contribution to the effort. It declares in bold letters that, "the twenty-first century will be the clean growth century." So Canada can "either act now — and take advantage of the global opportunity — or resign ourselves to being left behind."[30] An impression was being created that there is such a thing as a clean growth century and Canada could be leading the way into it.

"Clean Growth Century" moved quickly from the pages of the document into the public sphere. Coincident with the document's publication, the Twitter hashtag #cleangrowthcentury appeared. Several days before the prime minister and premiers signed the Pan-Canadian Framework document, an organization named Clean Growth Century was launched. Over a hundred environmental organizations, funders, alternative energy businesses, plus a sprinkling of faith and labour groups, signed on to an appeal to the first ministers to please get Canada into the clean growth century.[31]

The Clean Growth Century organization was initiated by a group of charities that are members of the Canadian Environmental Grantmakers Network — foundations that fund environmental work, broadly defined. They are also members of the network's low carbon funders group. This low-key group includes the Ivey, Metcalf, Catherine Donnelly, Tides, McConnell and North Growth foundations. Soon Environment Minister McKenna picked up the ball and ran with it. On her department website, in addresses to the Toronto Regional Board of Trade, Calgary Chamber of Commerce and Silicon Valley's Navigating the American Carbon World Conference, and on her Twitter account and Facebook page, she repeated the message: Taking action on climate change is a competitive advantage as we enter the clean growth century. But on the major action that needs to be taken — drastically cutting carbon dioxide emissions — not so much.

By combining their resources to promote a new kind of corporate growth — how new, really? — these foundations, established mostly by wealthy capitalists, act as gatekeepers against more radical kinds of solutions that might see capitalism itself as the true cause of global warming.[32] If voices calling for the radical restructuring of society don't get funded, they won't be heard.

For most foundations, the main source of their charitable giving is the interest and dividends they earn on the shares in companies they own. They are dependent on growth for their ability to provide grants to other organizations. Revenues will be greater if the companies they have invested in are growing. Clean growth is a natural. They can have their cake and eat it too — just change investments. They appear to be doing something valuable while ensuring that the system under which they operate continues to exist and prosper. At the Paris COP21, the European Climate Foundation, a major European philanthropic initiative, opined "the climate community's activities prior to and at the COP helped to lay the basis for the outcome."[33] It was an outcome with no binding standards, no mention of Big Oil's activities and reliance on emissions trading, an outcome, in other words of great benefit to Big Oil.

The Ivey Foundation is a leader in the low-carbon funders group. The once powerful Ivey dynasty accumulated its wealth over four generations of lawyers and business executives in London, Ontario and Toronto. At the end of 2016 the foundation had assets of CAD$95 million, yielding some $6 million from interest and dividends, gains on the sale of investments and other gains. It gave out $2 million in grants that year. The largest grant was $1 million over three years to the Ecofiscal Commission, the pro-carbon pricing think tank embedded in McGill University. This organization leads a network of environmental economists that is pushing market pricing solutions into every nook and cranny of human-environment interactions. Such action correlates with Ivey's top desired outcome for its funding, "to advance the understanding and adoption of pricing as a mechanism to be used to internalize environmental costs across our economy."[34]

It wouldn't be off mission then for Ivey's small program committee to include the managing director of the International Emissions Trading Association, the pro-market lobbying group that wrote the Paris Agreement's Article 6 authorizing emissions trading. As Edouard Morena of the University of London Institute in Paris, referring to foundations that fund environmental non-governmental organizations, puts it, most "grant-making [at COP21] went to groups supporting a reformist agenda grounded on the idea that, given the right policies, environmental preservation and corporate-driven capitalism were mutually reinforcing."[35]

PHILANTHROCAPITALISTS TO THE RESCUE

Catherine McKenna, meanwhile, was trotting the globe and engaging in grand-sounding ventures in an apparent effort to downplay the government's looming failure on the emission reduction front. A big theme was coal — an easy target. In December 2017 she was in Paris at the One Planet Summit hosted by French President Emmanuel Macron to announce a Canada-France Climate Partnership "to promote carbon pricing, coal phase-out, sustainable development and emission reductions in the marine and aviation sectors."[36] While in Paris she announced Canada's commitment to a joint approach with the World Bank Group to support developing countries as they transition away from coal-fired electricity.[37]

And when she was in Bonn for COP23, McKenna launched the Powering Past Coal Alliance, together with UK Minister of State for Energy and Clean Growth, Claire Perry. They collected a ragtag band of governments, businesses and international organizations that committed to phase out traditional coal power.[38] The use of the word "traditional" means that the alliance would accept coal plants if they were accompanied by carbon capture and storage. Such a qualification acts as an impediment to moving fully away from coal. Companies can still use coal if they can persuade a government or financial institution to contribute to carbon capture and storage. Canada and the UK were joined by some European countries, small island states, the provinces of Alberta, B.C., Ontario and Quebec and the city of Vancouver. Engie, a corporate sponsor of COP21 in Paris, was the only large power producer member of the alliance as of 2018. Its membership in Powering Past Coal means it commits to eliminate its traditional coal-fired plants by 2030. But it has to do this under the Paris Agreement in any case, so nothing is gained except a further burnishing of its corporate image.

Powering Past Coal should be an easy move for Canada. Ontario closed all its plants by 2014, the Harper government made moves in that direction and Alberta will have phased out its plants by 2030. Canada's coal consumption for power generation dropped from 52 million tonnes in 2005 to 35 million in 2014. But Saskatchewan, with seven coal-fired power plants, Nova Scotia with four and New Brunswick with one, still use coal power and have yet to commit to a complete phase-out. There is still a way to go.

The alliance teamed up with Bloomberg Philanthropies, owned by billionaire Michael Bloomberg, "to inspire and inform further efforts to retire coal-fired power plants in a practical, sustainable and economically inclusive way."[39] The Bloomberg Family Foundation has assets of US$7.9 billion, making it one hundred times larger and perhaps that much more impactful than Canadian charities like Ivey. Bloomberg dispensed US$345 million in 2016. With all the money going into research, strategies and publicity to reduce coal use, that much less is available for what should be the next step toward a net-zero-emissions society, getting rid of natural gas. Where's the Powering Past Natural Gas Alliance? It doesn't exist because natural gas is a destination, not an obstacle that must be powered past on the road to a zero-carbon future.

McKenna hooked up with the billionaires again after Canada was selected as a partner by the Breakthrough Energy Coalition, an organization of twenty-one of the world's richest men, headed by Bill Gates (2017 net worth, US$98 billion).

The billionaires' mission is to invest US$1 billion to help countries shift to clean energy, as defined by them.[40] By 2050 the world will be using 50 per cent more energy than in 2018, Gates predicts.[41] So clean investment opportunities are staggering, even for billionaires. Breakthrough connects government-funded research with interested investors in a massive public-private collaboration. Canada signed on as a public partner. But will investment go to research that might be crucial for the long-term health of the planet or to research that might be profitable for the investor? Government-funded research scientists work on projects they think will help bring down greenhouse gas emissions. Investors pick those that look most profitable. That's what gets developed. Important but financially unattractive projects languish by the wayside. It's a stark demonstration of neoliberalism's impact. The market, and not public policy, decides our future, just as Friedrich Hayek envisioned.

THE BUSINESS COUNCIL OF CANADA'S VISION FOR THE FUTURE

Despite significant opposition from some provincial governments, First Nations and environmentalists, the Business Council of Canada seemed to have succeeded in making its agenda the national agenda. And yet it was virtually invisible as Justin Trudeau, Rachel Notley, Murray Edwards, the low carbon funders group and the other players pushed forward on the various fronts. From its earliest days in the 1970s (as the Business Council on National Issues), the organization has shown an uncanny ability to direct government policies in directions desired by its members. Was it happening again?

John Manley, meanwhile, was strengthening his connections to Big Oil. In 2014, he was elected chairman of the CIBC board. The following year *Global Finance* magazine named CIBC the world's best investment bank in the oil and gas sector, with clients such as Suncor Energy, EOG Resources, Pembina Pipeline Corporation and Husky Energy.[42]

As for what's next on the Business Council of Canada's climate file, there's the February 2017 report, "Canada's oil sands: A vital national asset." In this publication, the CEOs say that regardless of gains made in renewable energy, "oil will be the main source of transportation fuel" until at least 2040. (Never mind Canada's 2030 goals.) Canada is "uniquely positioned" to provide this oil to domestic and international markets. The report reminds its audience that Canada has the third-largest oil reserves in the world, almost all of it oil sands. Most oil Canada exports goes to the US so it is important to find new markets. The report makes a pitch for pipelines to ensure oil exports can reach Asian and global markets.

We don't need to worry about climate change from our exploitation of this enormous resource, the report assures us, because greenhouse gas emissions from oil sands accounts for just 0.13 per cent of global emissions. And Alberta oil sands oil is cleaner than comparable oil from California, Mexico and Venezuela. Oil sands operations don't disturb the land, it alleges, nor do they use excessive water. Even better, pipeline and train transportation are safe and have minimal impact on people and the environment. And oil sands companies are working hard to make it even better.[43] Among the many distortions in this document, the claim that train transportation is safe would certainly come as a surprise to residents of Lac-Mégantic, Quebec.

CHAPTER 14
TWO DEGREES TOO LATE
(A SHELL GAME)

When NASA scientist James Hansen alerted the US Congress to the impending danger posed by global warming in 1988, the concentration of carbon dioxide in the earth's atmosphere was 353 parts per million (PPM). It had recently passed 350 ppm, the upper limit scientists consider to be safe in order to avoid dangerous climate change.

But Congress didn't listen. Twenty-seven years later, when President Barack Obama cancelled the Keystone XL pipeline, carbon dioxide concentrations reached 399 ppm. By the time Obama's presidency ended at the beginning of 2017, carbon dioxide emissions hit a record 403 ppm, a level not seen since the middle Pliocene era, three million years earlier. (The year 1750 is taken as the beginning of the modern carbon era, when carbon dioxide levels were about 280 ppm, and industry began to burn coal, oil and gas.)[1]

Four hundred ppm is not in itself a trigger at which time atmospheric conditions might turn deadly, but it is "a stark reminder that the world is still not on a track to limit carbon dioxide emissions and therefore climate impacts," warns Annmarie Eldering, deputy project scientist for NASA's Orbiting Carbon Observatory-2 satellite mission, which measures carbon dioxide emissions from space.[2] Most worrying to scientists was that the increase in 2016 "was the largest . . . we have ever seen in the thirty years we have had a network [of research stations around the world]," Oksana Tarasova, chief of the World Meteorological Organization's global atmosphere watch station, told BBC News.[3]

A second long-lived greenhouse gas is methane, which contributes nearly 20 per cent of total global warming. Currently methane levels are about two-and-a-half times higher than in the pre-industrial era, due largely to human activities like cattle breeding, rice agriculture, fossil fuel exploitation, landfills and biomass burning.[4] Methane emissions from these activities will need to be cut back to pre-industrial levels, just as for carbon dioxide.

 If greenhouse gas emissions continue growing at these rates, levels will hit 500 ppm within a half-century, with unknown impacts.

This may not happen, though, because at the Paris conference the nations of the world finally began grappling with ways to limit and then reduce emissions, albeit more than a quarter-century after James Hansen raised the alarm. What's more, for a brief period there was hope that emissions were already falling. Between 2014 and 2016, total global emissions due to energy use remained flat, even though economic growth continued, the result largely of reduced coal burning in China.

These hopes were dashed in 2017, when preliminary figures for the year projected that global carbon dioxide emissions would be up about 2 per cent, largely because China reversed its downward trend and emitted more pollution. "We hoped that we had turned the corner . . . We haven't," said Rob Jackson, an earth scientist at Stanford University and a co-author of a report by the Global Carbon Project, a team of seventy-six scientists from around the world that undertakes the annual study.[5] A note of caution should be attached to blaming the emissions increase on China. If the coal is used to create electricity to power factories that makes Apple iPhones for American and Canadian markets, who is really responsible for the emissions?

But even if emissions remain steady and even slightly decrease, they are not being reduced quickly enough. They need to get back to 350 ppm "if humanity wishes to preserve a planet similar to that on which civilization developed and to which life on Earth is adapted," Hansen warned.[6] The

Stockholm Resilience Centre and 170 other climate scientists agreed, urging the world to go back to 350 ppm to regain a safe level of risk.[7] There's even an activist organization, 350.org., that is named after 350 parts per million, "the safe concentration of carbon dioxide in the atmosphere."[8]

IS TWO DEGREES A SAFE LIMIT?

If 350 parts per million is one important number in climate change policy, then two degrees Celsius is another — the increase in temperature beyond pre-industrial levels that must form the upper limit to global warming to avoid "catastrophic" impacts to the planet. By the time of the Paris Agreement, in which two degrees was cast in stone, the planet was already halfway there. The year 2016 was recorded as the hottest since records started being kept in 1880, breaking the record set the year before, which in turn broke the record from the year before that. In total, the planet was about 1.1 degree Celsius warmer than pre-industrial levels. Such a level was beginning to approach Pliocene temperatures that were three or four degrees warmer than today. During the Pliocene, North and South poles were as much as ten degrees warmer, and sea levels at least fifteen to twenty-five metres higher than they are today, a terrifying prospect.[9]

The relationship between levels of carbon dioxide and temperature has long been a subject of intense research and speculation among scientists, as Chapter 2 discusses.[10] Two degrees continued to be discussed and debated during the decade leading up to James Hansen's testimony before the US Congress. Hansen concluded that the earth was warmer in 1988 than at any time since temperatures could be measured on a broad scale; that global warming had become large enough that we can be confident of a cause-and-effect relationship to greenhouse gases; and that the greenhouse effect was already influencing extreme events like summer heat waves.[11]

But Hansen didn't propose a limit to earth's warming that we must not exceed. That task fell to researchers at the Stockholm Environment Institute. Their 1990 report concluded that to avoid the worst impacts of climate change the safest option would be a limit of one degree. Since this had become impossible because of the volume of greenhouse gases already emitted, two degrees would be the next best limit. But, the report warned, there is nothing necessarily safe about two degrees. "Temperature increases beyond 1.0 degrees Celsius may elicit rapid, unpredictable, and non-linear responses that could lead to extensive ecosystem damage."[12]

The United Nations Framework Convention on Climate Change at the Earth Summit in Rio two years later did commit member countries to stabilize "greenhouse gas concentrations in the atmosphere at a level that would prevent dangerous anthropogenic interference with the climate system," but did not define this level.[13] The Kyoto Protocol, which followed five years later, also didn't define greenhouse gas and temperature limits. It asked signatory countries to cut emissions by 5 per cent below 1990 levels between 2008 and 2012.

During this period, there was increasing concern that the climate system might encounter catastrophic and nonlinear changes, a view popularized by journalist Malcolm Gladwell's best-selling book, *The Tipping Point*.[14] Little changes can have big effects and they can happen in dramatic fashion, was his book's message. Gladwell dealt with social phenomena; his concern seemed to be to understand how marketers can persuade consumers into buying new products. The idea was soon applied to natural processes. It was as if Gladwell's marketers turned their guns on climate policy makers and the public to persuade them to take two degrees of warming as gospel. Once a limit is surpassed, then abrupt and irreversible changes can be triggered, became a dominant climate change narrative. In the physical world, however, the consequence of crossing a critical threshold could take centuries to play out.[15] Major climate risk factors were identified long before Gladwell's book. Branding them as tipping points infused them with a sense of urgency.

Fear of abrupt climate change drove the political acceptance of a defined temperature limit. The two-degree limit moved into the political world when it was formally adopted by the Council of the European Union in 1996 and later by the G8. In 2010, participating nations in the UN Framework Convention on Climate Change signed the Cancun Agreements, finally committing them to a goal to limit average global temperature warming below two degrees in comparison to pre-industrial levels.[16] In 2015 in Paris, negotiators adopted two degrees as the upper limit, with an aspirational goal to limit warming to 1.5 degrees.

They were twenty-five years too late. By 2017 the possibility of limiting warming to two degrees was fast receding. One often cited study in the journal *Nature Climate Change* found a likely range of global temperature increase by 2100 to be between 2.0 and 4.9 degrees, with a 5 per cent chance of limiting warming to two degrees and only one chance in 100 of keeping man-made global warming to 1.5 degrees.[17] And this may be optimistic. Even if fossil fuel emissions were to suddenly cease, because of past emissions as much as 1.5 degrees of warming may already be locked in.[18]

There's a third qualification to the Paris two-degree limit. The parties agreed to hold the temperature increase to "well below 2 degrees Celsius above pre-industrial levels." But it doesn't define pre-industrial. The Fifth Assessment Report of the Intergovernmental Panel on Climate Change, published in 2014, takes 1850–1900 as its base, but this is not pre-industrial. The industrial revolution began around 1750 and the concomitant growth in greenhouse gas emissions was well underway by the late eighteenth century. If the earlier date of 1750 is taken as the pre-industrial marker, then another 0.2 degrees must be added to current temperature increase.[19]

THE CARBON BUDGET

Along with 350 parts per million and two degrees Celsius, a third number is important, but even less certain: 600 billion. This is the tonnes of carbon dioxide that can be emitted if the world is to have a reasonable chance of keeping warming considerably below two degrees. It's called the carbon budget. However, it could be 800 billion or even more. The concept is clear, but the calculation is convoluted, influenced by myriad factors.

Calculating the carbon budget for the 1.5-degree target provides an illuminating example. The UN's Intergovernmental Panel on Climate Change 2014 report calculated a budget of 400 billion tonnes to be the maximum amount of carbon dioxide humanity could emit into the atmosphere after 2011 and still have a 66 per cent chance of staying below 1.5 degrees. By 2015, with emissions at just over 39 billion tonnes a year, the remaining budget number had dropped to about 245 billion tonnes which would be used up in just over six years.[20] In 2017, however, a new analysis published in the journal *Nature Geoscience* estimated the remaining budget based on 2015 to be closer to 880 billion tonnes, and over twenty years at current emissions until the 1.5-degree limit is breached.[21]

The study did raise some cautions. For one thing, it estimates human-caused warming in 2015 to be 0.93 degrees above pre-industrial levels. As mentioned earlier, other studies put the level at 1.1 degrees, which is considerably closer to 1.5. Whatever the exact number, humanity is not being let off the hook for a few more decades, before it must act. Every model requires that emissions "start dropping off a cliff in short order, hitting zero by 2080," cautions American science journalist Scott Johnson.[22]

The carbon budget has an important consequence — it determines how much of the fossil fuel in existing coal mines, oil wells, oil sands and gas fields around the world can be burned before the budget is exhausted. That's why there's so much debate over the budget. Big Oil has a lot at stake. A 2016 study by Oil Change International, a Washington, D.C.-based research and advocacy organization, found that burning the fossil fuels in currently operating fields and mines, if they run to the end of their projected lifetimes, will far surpass 1.5 degrees and even exceed the two-degree limit.

In November 2017, 15,000 scientists from 184 countries assessed the world's latest responses to various environmental threats. Global climate change sat in number-one spot on the list that included declining access to fresh water, forests and fisheries. Scientists had written a similar warning to humanity twenty-five years earlier at the time of the Rio Conference. They said then that humans had pushed Earth's ecosystems to their breaking point and were on the way to ruining the planet. The 2017 letter was even bleaker: "Humanity has failed to make sufficient progress in generally solving these foreseen environmental challenges, and alarmingly, most of them are getting far worse," they wrote.[23]

As if to add to the pessimism in the scientists' notice to humanity, the day after they released it, the International Energy Agency issued its *World Energy Outlook* for 2017.[24] In this publication the Paris-based organization predicts in its New Policies Scenario that global energy demand will rise by 30 per cent by 2040. This number is based on the most recent policy directions reported by all countries.

The good news is that renewable sources of energy will account for 40 per cent of the increase. But that leaves oil, gas and coal to account for the other 60 per cent. Oil demand will grow by 8.5 per cent, and natural gas demand will rise by 45 per cent. Coal's boom years are over, the *Outlook* predicts, but coal will still add 400 gigawatts of power generation capacity.

The future envisioned by the International Energy Agency is far removed from what the scientific community claims is necessary for a liveable planet. But scientists are no longer in charge of defining what kind of a threat global warming is to humanity. That crucial task has been captured by Big Oil.

SHELL'S MANY MEANS OF AVOIDING CHANGE

In this context, just as in the context of Canadian vs. foreign ownership in the oil patch, in Chapter 12, Shell makes a good illustration. No matter whether the subject is emission targets or renewables or carbon capture or liquefied natural gas, Shell has provided an ingenious kaleidoscope of reasons it cannot stop growing its carbon footprint.

Yet to the public Shell presented itself as an energy innovator, inviting social-media input to figure out the future via a rock video-infomercial. Featuring an international cast of six artists led by Jennifer Hudson, "Best Day of My Life" ("fun and sticky" — *Adweek*) was the most-watched ad on YouTube in December 2016, with more than 24 million views.

In a speech at an OPEC meeting in Vienna in June 2015, Royal Dutch Shell group CEO Ben van Beurden laid out the company's vision of its energy future. "The energy transition is not about one system replacing another," he cautioned. "It's about finding a way of integrating the old and the new systems. The two evolving alongside each other — and complementing each other." So forget about phasing out oil. Shell will continue to explore for it, produce it and sell it, while adding natural gas, some renewables and carbon capture and storage if someone else will pay for it.[25]

It was an approach that John Ashton, the UK's former special envoy for climate change, called "narcissistic, paranoid and psychopathic."[26] Ashton challenged van Beurden on his vision of the energy future — a so-called "energy transition," but one without any "change 'in the longer term' in the drivers of supply and demand for oil." The issue for Shell, Ashton claimed, "is not how to deal with climate change but how to do so without touching [its] business model."[27]

It wasn't until the Paris Agreement, with its lack of mandatory limits and its promotion of markets, was in the bag that Shell and other companies started to commit to doing something. Shell was under growing pressure from shareholders to take some kind of action. But whatever it did, Shell expected to be able to pump all the oil and gas reserves it had listed on its balance sheet. It wouldn't have to strand any assets. "The company is valued on reserves that we can produce in the next twelve or thirteen years," CEO van Beurden said in a newspaper interview. "We should certainly be able to produce those under any climate outcome. Even if global temperatures can only rise by two degrees."[28]

But as the company used up its reserves, it was actively replacing them. Shell was bolstering its position in natural gas with the US$52 billion takeover of BG Group, the former state-owned British Gas. "We're more a gas company than an oil company," van Beurden told *Bloomberg News*. "If you have to place bets, which we have to, I'd rather place them there."[29] Shell Canada country chair Michael Crothers explained why in a speech at the 2016 Globe Conference in Vancouver. "Now don't get me wrong," he told 2,000-plus delegates from fifty countries. "I understand that although it emits half the carbon dioxide of coal in power generation, natural gas doesn't come without its critics. But if global decarbonisation is the prize and if we're willing to be balanced in the energy debate, then I believe natural gas is part of the energy transition."[30] Well perhaps he didn't explain it at all.

While boosting natural gas production may work as a strategy for the oil company, it won't do much to create the low- or zero-carbon economy that's required. Switching from coal to natural gas to create electricity is a no-brainer because it cuts carbon dioxide emissions by nearly 50 per cent. It's an easy argument for Shell to make because Shell no longer does coal.

But switching from oil to gas is not nearly as impactful, since it reduces emissions by just 25 per cent per unit of energy burned. Gas-fired plants still emit sulphur dioxide and nitrogen oxides that contribute to acid rain and ground-level ozone that can damage forests and agricultural crops. There's also the problem that most natural gas in North America is produced through fracking, with its attendant health and environmental risks.[31] Shell may argue that natural gas is a transition fuel on the road to the low-carbon future, but it will be a long transition indeed because of the costly pipelines and liquefied natural gas facilities that need to be built and that will require decades for their costs to be amortized. Gas is "not just going to be a bridge," van Beurden admitted to *Bloomberg News*, but a lucrative part of the energy mix indefinitely.[32] So forget about a low-carbon future. "Natural gas is not a solution for climate change," David Suzuki accurately points out.[33] Nonetheless, it's Shell's solution. And by investing heavily in natural gas, the company ensures fewer resources will be available for renewables, which in contrast to natural gas are the only path to a low-carbon future.

SHELL'S HALF-HEARTED INVESTMENT IN RENEWABLES

Shell did move on the renewables front after Paris by setting up a New Energies business, deploying US$1.75 billion in the first year and a plan to spend $200 million a year to explore and develop "new energy opportunities." The investments were minute given the US$25 billion a year the company was set to spend each year. Shell's annual report emphasizes that the company will invest in renewables only "where sufficient commercial value is available." Later in 2017 van Beurden upped new-energies annual spending to US$2 billion, a ten-fold increase over his original commitment, but still less than 10 per cent of total investment spending.[34]

The company would invest in commercially viable wind and solar projects, but only in conjunction with natural gas, so it can manage the "intermittency" that can be problematic with wind and solar sources of energy. Shell uses gas "as a partner with renewables to ensure steady power supplies when the sun does not shine or the wind does not blow," Shell's chairman explains.[35] This is a clever way to get an old energy source — natural gas — into the new energies portfolio. Shell has invested in power trading, but its presence in power storage has been minimal. If the ability to store renewable power became a priority for the company, its claim that it needs backup gas capacity would lose its salience. In fact, New Energies and Integrated Gas operations are overseen by the same Shell director. New Energies was likely a product not of any operating division but of Shell's vaunted marketing team, intended "to position Shell as a key player in the world's energy future," or at least to create that appearance.[36]

While Shell was repositioning its renewables portfolio, the European

Union (EU) was reaching a difficult agreement to set an overall target of 40 per cent greenhouse gas reduction below 1990 levels by 2030. Originally the EU proposed legally binding targets for individual member states on energy efficiency and renewables. But, like all efforts to impose mandatory limits, these did not make it into the final agreement. Documents released to the *Guardian* newspaper revealed that the binding targets were removed because of intensive, prolonged lobbying by Shell.

The oil giant, one of the most influential lobbyists in Europe, wants a market-led strategy of gas expansion and not legislated renewables levels. "Shell believes the EU should focus on reduction of greenhouse gases as the unique climate objective after 2023, and allow the market to identify the most cost efficient way to deliver this target, thus preserving competitiveness of industry, protecting employment and consumer buying power, to drive economic growth," Shell's Malcolm Brinded wrote to the European Commission president in one of the released documents.[37]

Let government set limits and then let the market do the rest. Such an approach does little to drive the energy transition to renewables: renewables if necessary, but not necessarily renewables. In Shell's world, renewables have their place, but given the globe's population trajectory over the coming decades, phasing out fossil fuels is not realistic, says CEO van Beurden.[38]

Freed from mandatory emission cuts, Shell pledged to reduce greenhouse gas emissions by 20 per cent by 2035 and by half by 2050. Shell would meet these targets largely through its transition to natural gas and not by leaving oil assets in the ground.[39]

A third Shell climate change strategy has been to develop carbon capture and storage projects so it does not have to reduce its exploration and production, especially for oil sands oil. Before it off-loaded its stake in the Athabasca Oil Sands Project, in 2015 Shell completed the $1.35 billion Quest project near Fort Saskatchewan, Alberta, to bury 30 million tonnes of compressed carbon dioxide two kilometres underground. The carbon dioxide was captured from Shell's bitumen upgrader at nearby Scotford and reduced the upgrader's emissions by about one-third, or one million tonnes a year.

Carbon capture and storage is an expensive strategy, and it works only if taxpayers contribute the lion's share of costs. In Quest's case, the federal and Alberta governments paid $865 million — nearly two-thirds of the project's cost, justifying their contributions on Shell's promise to share whatever it learned on the project. But the main lesson was that carbon capture and storage isn't feasible without taxpayer support. "It would be quite a challenge to build another carbon capture and storage plant without [taxpayer] support, of course," Shell country chair Lorraine Mitchelmore admitted during a conference call with journalists.[40]

An alternative to taxpayer support is a price on carbon, something that Shell promoted at COP21 and something Shell has supported since the corporate pivot to carbon pricing a decade earlier. Van Beurden figures a carbon price of between US$60 and $80 a tonne would mean, "companies like ourselves would feel compelled to capture and store the carbon dioxide rather than emit."[41] Shell's Quest and the other twenty-one fossil fuel industry plants operating or coming on stream by the end of 2018 will have a capacity of 40 million tonnes a year. But the International Energy Agency warns that carbon capture and storage would need to sequester about six billion tonnes by 2050 — or 150 times more — to rein in global warming. The cost could be upward of US$6 trillion.

A NET-ZERO EMISSIONS FUTURE? NOT FOR SHELL

Boosting natural gas production, continuing to search for and produce oil, and developing taxpayer-backed carbon capture and storage facilities will never achieve a net-zero carbon dioxide emissions goal and may even increase emissions. That inconvenient truth, however, doesn't seem to prevent Shell from positioning itself as a leader on route to a "net-zero emissions future" — a strategy that was revealed in a leaked Shell call for proposals from public relations firms to activate a scenario titled "A better life with a healthy planet: Pathways to net-zero emissions."[42]

The winning PR firm will have its work cut out because none of the pathways are routed through Shell country. The company admits in the document "while we seek to enhance our operations' average energy intensity . . . we have no immediate plans to move to a net-zero emissions portfolio over our investment horizon of 10–20 years. Net-zero emissions, as discussed in this document, is a collective ambition that is applied in the aggregate."[43]

Net-zero emissions for everyone else, but not for us.

The call for proposals identifies Canada as a primary target in the campaign. Within Canada, "energy-engaged millennials" are the tier-one targets, with their interest in energy or in related subjects such as social progress, urbanization, technology and innovation, although the document doesn't explain how these subjects are related to energy. Forget older consumers, the briefing document declares. They are too set in their ways. Go after impressionable younger people with whom we can develop brand loyalty. (This is exactly the audience that was targeted with "Best Day of My Life.")

This goal may help explain Shell Canada's Quest Climate Grant program, which awards $50,000 per project to "exceptional young entrepreneurs" with "bold new ideas and smart thinking" in tackling climate change.[44] Other targets in the net-zero emissions campaign are government officials, business leaders and major opinion shapers, such as editors and columnists, to "position Shell as a leader in helping societies grapple with future energy challenges."

It's a classic bait-and-switch strategy. Start with something we all want — a low carbon future. Divert attention away from something we all know — the low-carbon future can be achieved only by drastically cutting fossil fuel use. Substitute a different goal — such a future can be achieved through a "patchwork of solutions across different sectors," the call for proposals explains. Reinforce the new message — "if society only focuses on energy we'll miss the mark." It's Shell spin at its finest. And it would soon be reinforced by an old ally.

BIG OIL RECRUITS THINK TANK

In March 2018, B.C.'s New Democratic government announced it would provide substantial tax incentives to encourage LNG Canada, the consortium seeking to build a natural gas liquefaction and export facility at Kitimat.

Although "Canada" is in the enterprise's name and former premier Christy Clark acted extensively as its pitchwoman when she was in office, Canada is actually conspicuous by its absence in the consortium. And conspicuous by its presence is the Chinese state, in the form of the publicly traded arm of China National Petroleum Corporation, which sits at number 4 on the 2018 Fortune Global 500. Right next door at number 5 is its partner Shell. The other two founding participants are the Japanese giant Mitsubishi and a global liquefied natural gas powerhouse, Korea Gas Corporation.

Yet these titans of free enterprise evidently need subsidies. And a government comprised of social democratic and environmental parties obligingly offered them, exempting LNG Canada (and other liquefied natural gas projects) from provincial sales tax for the construction of the facility, reducing the carbon tax on the plant's operation and scrapping an income tax surcharge that had been levied by the previous Liberal government. The B.C. government would be foregoing $6 billion in potential revenues — about 20 per cent of what it would have otherwise received from the project over 40 years, or about $150 million a year.

The news was of course, welcomed by the company. Susannah Pierce, LNG Canada's director of external relations, said the "measures are very important . . . and they're very timely."[45] The timing was right. It seemed to give the signal Royal Dutch Shell and its Asian partners were waiting for on the road to making a final investment decision. Within a month, the consortium named a main contractor for the project. And within two months, it would have a fifth partner: Malaysia's Petronas, which had scrapped its own CAD$11.4 billion liquefied natural gas joint venture on Lelu Island south of Price Rupert, B.C., was taking a 25 per cent stake in the project.[46] Petronas held one of the largest natural gas reserves in the Montney formation, which spans a vast area from Edson in western Alberta to Fort Nelson in northeastern B.C.

The NDP incentives were also welcomed by the Fraser Institute, an odd position to be taken by an organization normally hostile to any government interference in the market. The institute's top guns, president Niels Veldhuis and senior director of natural resource studies Kenneth Green, wrote an op-ed in the *Vancouver Sun* celebrating the government's actions as "a bright spark of light in an otherwise gloomy environment for energy transport and export infrastructure."[47] Natural gas was a cleaner option, they wrote, but they did not indicate cleaner than what. And Asian customers were eager to import Canadian gas, the two claimed, because Canada was a "reliable, democratic government with some of the highest environmental standards in the world." (Is this really why state-owned companies like Petronas want Canada's liquefied natural gas? And as for investor-owned companies like Royal Dutch Shell, in the quest to obtain the oil and gas it needs, it has been dealing with dictators and despoiling environments for over a century.) And better yet, they concluded, thousands of construction jobs and hundreds of operational jobs would be created. It was good news all around.

The Fraser's message must have been particularly welcome to LNG Canada's Pierce, a long-time Shell public relations official. Of course, it's not that she didn't know about the institute's work; she had recently been appointed to its board of directors. The op-ed could be an early product of a renewed collaboration — Shell had been a donor in earlier years. Pierce was joining eight other fossil fuel industry executives on the Fraser board.

Why would so many oil patch people congregate here? They covered all segments of the industry, from exploration and development to venture capital and investment to pipeline transportation to marketing and liquefied natural gas production. The most prominent member of this contingent was Gwyn Morgan, former CEO of Encana, who built that company into a major presence in B.C. fracking operations and advised Liberal premiers Gordon Campbell and Christy Clark on how to exploit the resource.

Neither Morgan nor any other directors were experts in the economics or science subjects institute staff and fellows research. So why were they there? One answer might be that they put money into institute work. Morgan donated CAD$1 million to the think tank through his family foundation and served as a vice-chairman of the board for several years after that, indicating perhaps, a relationship between the size of one's donation and whether one even gets to be on the board. As to what the money might be buying, former executive director Michael Walker long maintained that institute directors could not tell the staff what to research, let alone dictate research findings.

FRACKING IS NOT A PROBLEM

The Fraser Institute's Tobacco Papers, discussed in Chapter 5, provide a different interpretation of how think tank funding and research intersect: The organization comes up with a research program and then seeks out corporations that will benefit from the research to finance the work.[48]

Green had returned from the American Enterprise Institute in 2009 to lead the new Fraser Institute centre located in Calgary, perhaps to be near the sources that were supporting the initiative. In any case, that would explain the motivation for a 2014 Green study claiming that risks due to fracking for oil and gas are "modest and manageable with existing technologies."[49] In the study, "Managing the risks of hydraulic fracturing," Green lists major Canadian unconventional natural gas formations and the contribution they could make to the Canadian economy. He makes much of the potential of the Montney formation, whose "marketable unconventional gas resource is one of the largest in the world."[50]

Six months before the study was released, the institute added a new fossil fuel director who happened to have multi-million-dollar holdings in fracking companies — Calgary's Ron Poelzer, a director of NuVista Energy, which has major Montney operations. Green's diagnosis of "modest" risk would certainly help the company's prospects and Poelzer's own pocketbook, as he owned CAD$28.2 million worth of NuVista shares and options in 2018.[51] It would also help Royal Dutch Shell, LNG Canada's major partner, which has large holdings in the Montney and Duvernay plays, where it would extract the gas to feed the Kitimat plant.

In his study, Green relies on a recent evaluation of fracking risks by the Council of Canadian Academies, a research organization supported by Canada's top scientific societies, established in 2005 to assess science relevant to public policy issues. Fracking is certainly one of these.[52] The blue-ribbon panel of scientists notes, "the health and social impacts of shale gas development have not been well studied," while fracking may "adversely affect water and air quality and community well-being."[53] The study concludes, "despite a number of accidents and incidents, the extent and significance of environmental damage is difficult to evaluate because the necessary research and monitoring have not been done."[54]

Somehow Green gets this all wrong. His conclusion that risks from fracking are "modest" is a distortion of the very scientific literature he cites, which is unanimous in its findings that the magnitude of risks is unknown because of the lack of information. Green's "modest" is not the same as the scientific community's "not enough is known." The Council of Canadian

Academies scientists seem to be applying the precautionary principle to fracking. This principle was enunciated by a conference of scientists, lawyers, policy-makers and environmentalists in 1998 at the headquarters of the Johnson Foundation in Racine, Wisconsin. It states, "when an activity raises threats of harm to the environment or human health, precautionary measures should be taken even if some cause and effect relationships are not fully established scientifically."[55] The proponent of an activity and not the public, should bear the burden of proof, the principle continues.

The Council of Canadian Academies provided a checklist of areas where precaution should be applied, where the industry needs to prove itself before bans are lifted; Green garbles these. When the council says, "it is not clear that there are technological solutions to address all of the relevant risks,"[56] Green claims that "risks . . . can seemingly be managed (but not entirely eliminated) with existing technology."[57] When the council says, "full disclosure of chemicals and the chemical composition of flowback water is . . . necessary,"[58] Green provides a table listing the types of chemicals that may be used in fracking, but does not discuss the impacts these chemicals can have on ground water, human health or the environment.[59] It's a table without any meaning.

Fracking is not a concern, he says, because there has been "perhaps only one documented case of direct groundwater pollution resulting from injection of hydraulic fracturing chemicals used for shale gas extraction." However, the source he quotes[60] was referring to an Environmental Protection Agency report to the US Congress in 1987,[61] decades before fracking became the potential threat it is today. Green uses his study to attack bans on fracking, claiming "there's no evidence of unmanageable risk associated with hydraulic fracturing that justifies a ban or moratorium."[62] It's as if the Council of Canadian Academies study had never happened.

Green produced an update the following year, reinforcing his earlier study's findings: "Research on the safety of hydraulic fracturing confirms that while there are indeed risks with it, they are for the most part readily manageable with available technologies and best practices."[63] In the process, though, he misstates the name of his primary source calling it the Canadian Council of Academies throughout his report.

Distorting and misquoting scientific and economic research has been a

Fraser Institute mainstay. This brings to mind Philip Mirowski's warning about the uneasy relationship between neoliberalism and the truth. For neoliberal think tanks, he says, lying and spreading falsehoods about global warming is as valid as messages disseminated by scientific experts "because the final arbiter of truth is the market and not some clutch of experts who represent sanctioned science."[64]

Careful studies of Fraser Institute research have documented this phenomenon. A telling illustration of Fraser Institute methodology is what environmental scholars Hilda McKenzie and Bill Rees call "crackpot rigour," an apparent precision that results from fundamental and ideological distortions of the underlying statistics and data used to construct indexes. The claim in one report that overall environmental problems in the United States were 19 per cent less severe in 1995 than in 1980, and in Canada 17 per cent less severe, McKenzie and Rees say, is an example of such crackpot rigour.[65]

A 2017 evaluation of Canada's air quality since 1970 continues the crackpot-rigour tradition. Subtitled "An environmental success story," it concludes that "air quality in Canada has improved substantially . . . [occurring] . . . at the same time as considerable growth in Canada's population, economic activity, energy use and consumption of motor fuel."[66] However, the report doesn't compare Canada's record with that of other countries; it doesn't credit the role of government in driving down levels of pollutants; and, most important, it doesn't even mention the pollutant that most concerns people today, carbon dioxide. The authors conclude that air pollution has been decoupled from energy use and population growth. Therefore we don't need to tighten emissions policies. Once again, the think tank's crackpot rigour is utilized to defend against government regulation. It's a rigour that Big Oil, evidently, finds to its liking.

CHAPTER 15
THE CORPORATION VS. NATURE

The world may never recover from neoliberalism's biggest hoax, the assertion that global warming is not occurring — or, if it is, it's beneficial. The phalanx of fossil fuel industry-funded deniers has variously claimed that: global warming isn't occurring, it's global cooling; if there is global warming, it's due to sunspot activity; or global warming is a plot concocted by Al Gore, the world's climate scientists and the United Nations to take away our freedoms and boost big government; among other canards.

The sheer weight of scientific evidence and people's ever-more-frequent experiences with extreme weather events have muted, but not silenced, these narratives. A significant minority of Americans — including some in high places — and of Canadians in the three western provinces, profess to believe that global warming is a lie. These non-believers provide a receptive audience for neoliberal think tanks that still promote global warming denial, such as the Heartland Institute in the US, and the Frontier Centre for Public Policy in the western provinces. The work of these think tanks makes concerted action on global warming more difficult to accomplish, allowing the fossil fuel industry to reap its gargantuan profits for several more decades.

Nonetheless, much of the world has moved on, understanding the dangers of climate change and the need for urgent action. But we have run headlong into a second neoliberal hoax, the claim that we need to simply bring in a carbon tax and switch to clean energy, and technology and human ingenuity will enable us to carry on much as before. It's a comforting myth. The debate is over how quickly this must be done. We can then leave it to the market to deal with any problems and we will all be more prosperous.

This claim that clean technology will lead to a brighter future is the latest example in Canadian policy-making of technological determinism, the belief that developments in technology cause changes in society, changes we cannot resist. One of our most enduring beliefs is that Canada has survived as a nation and resisted the expansionist urges of our neighbour to the south by building east-west communication links. It's a myth because many communication links — telegraph, railway, communication networks, even pipelines — run north-south as often as they run east-west. Canada survives despite these communication networks.[1]

The federal government's Pan-Canadian Framework on Clean Growth and Climate Change informs us that "clean technology [enables] the sustainable development of Canada's energy and resource sectors, including getting these resources to market."[2] That's quite a burden to be placed on clean technology's shoulders. The government must mean it, though. The phrase "clean technology" is used fifty-nine times in the eighty-six-page document. It brings to mind "If you build it, they will come" in the movie *Field of Dreams* — and is just as likely to happen in the real world.

The rest of this chapter looks at four pathways back to the real world. We can question growth. We can question economists. We can listen to Indigenous voices. And we can listen to Nature.

WE CAN QUESTION GROWTH

A continued emphasis on growth and consumption, albeit with a different mix of energy sources, is what Big Oil has worked for since 2007 when industry pivoted from denial and voluntary measures to clean growth and carbon pricing. We've seen how Big Oil and its corporate and political allies went about ensuring mandatory emission levels were sidelined in favour of vague commitments to carbon pricing, clean technology and voluntary approaches. These successes — even the Paris Agreement contains no mandatory emissions reductions — left the door wide open for oil companies to determine their own futures.

Working quietly behind the scenes and in high-profile organizations like the Business Council of Canada and the World Business Council for Sustainable Development, Big Oil moved the debate away from discussion of the drastic measures required to create a zero-carbon world and toward market-based "solutions" that may cut carbon dioxide emissions, but not enough to prevent severe climate impacts.

In fact, Royal Dutch Shell and the other successors to the Seven Sisters are well on the way to transitioning from Big Oil to Big Energy, as they spend billions of dollars acquiring renewable energy assets, while continuing to spend on oil and gas development. By whatever means, the future means growth, since this is critical for investors. Being some of the largest companies on the planet, they'll end up owning the renewables sector. But if Shell must grow to survive, it doesn't follow that society must do the same.

Of course, clean growth is better than dirty growth. For 150 years, the developed world has been mired in the muck of burning coal, oil and gas, spewing noxious carbon dioxide, methane and other polluting substances into the atmosphere. And the developing world is contributing its share of greenhouse gases, in part to produce goods for consumption in the developed world, and in part to emulate the Western big-footprint lifestyle. So the switch to non-polluting sources of energy is positive. But if giant multinational corporations continue to wield outsized power over global affairs, and if neoliberalism remains the dominant ideology, then not much will really change.

Maurice Strong led the effort to make climate change a global problem, disempowering local communities in the process. At the Paris talks, continued growth and globalization remained the overarching context for the non-solution that was reached. No mention was made about the possibility that promoting diversified local economies may be a more effective way to fight climate change.

The neoliberal way forward is an economy much like what exists in 2018, except for some change in energy sources. Species will continue to disappear, ecosystems will continue to deteriorate, and inequality will continue to increase. How much of the earth's surface can be covered with wind and solar-panel farms without further displacing endangered species and Indigenous peoples? What are the limits to hydro power, especially given forecasts of declining water resources? And how clean are massive hydropower projects? What other resource limits will we soon encounter?

"Deploying renewables on the scale that is required to address climate change demands enormous quantities of concrete, steel, glass and rare earth minerals, and vast programs of land-clearing to house solar and wind plants," writes Australian social and political theorist Ben Glasson.[3] Governments trapped in neoliberal thinking are unable or unwilling to take the strong measures necessary to cut carbon use, let alone address these other pressing issues.

Clean energy and clean growth advocates appear to have no plans for how to lessen inequality.[4] The requirement set by many economists that carbon pricing must be revenue neutral removes one powerful weapon for creating a more equal world — use revenues from global carbon pricing to reduce poverty.[5] If economic disparities have been driven by an oil-fuelled economy, how will renewables lessen inequality? What about child poverty, homelessness, affordability?

Nor does the clean-energy movement seem to have plans for preventing species extinction. As ecologist Bill Rees puts it, "The competitive displacement of other species is an inevitable by-product of continuous growth on a finite planet."[6] Clean production and consumption must translate into less production and consumption by fewer people, Rees says. Instead of growth, we need cooperative redistribution. Clean energy must mean less energy; clean growth must mean less growth. Neither the fossil fuel industry nor advocates for clean growth in government or the non-profit sector are talking about slower growth or no growth.

Neoliberalism has hobbled our ability to conceive of solutions that could have a real impact. The energy future and its accompanying economic, social, cultural and ecological transformations cannot be imagined in 2018, but some ingredients are clear. The planet cannot accommodate unlimited economic growth ad infinitum. There must be limits to growth and greater emphasis on redistribution. And the energy future, regardless of how clean it is claimed to be, must be based on an understanding of, and respect for, ecosystems. Without healthy ecosystems, the planet will not survive in a form we recognize, even if humanity manages to keep global warming within two degrees.

WE CAN QUESTION ECONOMISTS

David Suzuki, Canada's most trusted environmentalist, has little time for economists. "Conventional economics is a form of brain damage," he cautioned in a 2011 documentary, *Surviving Progress*. "Economics is so fundamentally disconnected from the real world, it's destructive."[7] Suzuki's main concern is how economists use the concept of externalities. Carbon pollution is the ultimate externality for which producers and consumers of fossil fuels do not pay the full price. They should. But Suzuki sees a fundamental flaw in the concept. In an interview in *Common Ground* magazine Suzuki explains why:

> *Nature and nature's services — cleansing, filtering water, creating the atmosphere, taking carbon out of the air, putting oxygen back in, preventing erosion, pollinating flowering plants — perform dozens of services to keep the planet happening. But economists call this an externality. What that means is "We don't give a shit," It's not economic.*[8]

These essential services, or better, functions (because services implies services for humans whereas functions implies functions of nature), are external to the human economy at least until humans figure out how to monetize them. Then they are internalized within the economy and people can make money from them. This is not a viable option for a truly sustainable future but it's the environmental economists' option. The biosphere is what keeps us alive. The economy is a human construct. Yet we elevate it above nature, Suzuki explained.[9] "As long as economic considerations trump all other factors in our decisions, we'll never work our way out of the problems we've created."

Of course economists bristled at the insults Suzuki hurled at them. "He displays a perplexing lack of understanding of basic economic concepts . . . and owes economists an apology," one harrumphed.[10] How could "Canada's most famous environmentalist . . . get the most important concept in environmental economics so badly wrong," another wondered.[11]

The harshest reaction came from University of Alberta economist Andrew Leach, who authored Rachel Notley's climate action plan that endorsed increased oil sands production and set a price on emissions. Suzuki was set to receive an honorary degree from the university, a decision that was widely denounced in the province because of his anti-oil sands politics. Leach would not share the stage with Suzuki, he wrote in a Twitter thread, not because of Suzuki's politics but because he had defamed the economics profession by calling it "a form of brain damage" and a "pretend science."[12]

But how far off the mark is Suzuki? Pricing externalities is the economics profession's main contribution to fighting climate change. And it has captured the high ground in the fight — other options have been marginalized. This is diverting attention away from more effective actions that might end up having greater impact on corporate profits, such as mandatory emission ceilings and stringent regulations.

If economists really believe that externalities identify costs not reflected in prices, wonders Bill Rees, why aren't they actually identifying externalities associated with land and soil degradation, deforestation, biodiversity loss, mining, oil and gas development, pesticide use and other damaging externalities?

Economists seem all but absent from these battles, he notes. Correcting prices for all these failures would likely exceed the associated benefits.[13] If all externalities are accounted for, growth will be uneconomic and will make the world poorer. No wonder Suzuki calls growth-directed economics "suicidal." "Economists think that even though we live in a finite biosphere, the economy can grow forever. It can't."[14]

Shifting costs onto the environment, the community or future generations is the way profits are normally made in the modern business system. Imposing a carbon tax of $30 or $50 a tonne of carbon dioxide is mere tokenism. The true cost should be $200 to $300 a tonne, which would ruin the capitalist enterprise and stimulate the rise of a different system.

WE CAN LISTEN TO INDIGENOUS VOICES

American geochemist Dave Keeling first began measuring the amount of carbon dioxide in the atmosphere in the 1950s from his station at Mauna

Loa Observatory in Hawaii. Keeling is best known for his work documenting the relentless rise of carbon dioxide in the atmosphere over decades, as Chapter 14 describes.

But Keeling also found a persistent daily fluctuation in which the air contains more carbon dioxide at night than during the day, as plants and soil exhale it at night and inhale it during the day. It is as if the Earth and life on it is breathing, creating a powerful counter to the prevailing view that the Earth is mere real estate. Keeling also discovered a seasonal variation of about five parts per million, as nature withdraws carbon dioxide from the air for plant growth during summer and returns it each succeeding winter.[15] But the Earth, with all its ecosystems, is choking to death as Big Oil's products pump more and more toxic fumes into the atmosphere.

Indigenous peoples, whose laws and cultures arise from the land, are among the most vulnerable to climate change; they are among the staunchest defenders of the earth; and they have knowledge of local environments and conditions accumulated over generations.[16] In short, they have much to contribute to the fight to contain global warming and counter Big Oil's power. Recognizing their laws and legal orders may be humanity's best hope for championing ecosystem health and species survival.

Yet Indigenous peoples have been systematically excluded from official decision-making and policy-formulation in the United Nations Framework Convention on Climate Change. This is ironic, given that the 370 million Indigenous peoples would create the third most populous country in the world, behind China and India, but larger than the United States, with its 325 million people.

At COP21, references to the rights of Indigenous peoples were cut from the binding sections of the agreement, relegating the only mention of Indigenous rights to the non-binding aspirational preamble, writes Mitch Paquette of Intercontinental Cry, an Indigenous news organization.[17]

In the preamble, the "rights of Indigenous peoples" is included in a list of twelve things parties "should" consider when taking action on climate change. Other items on this list are migrants, gender equality and the right to health. In short, it is a shopping list. The use of the word "should" rather than the more directive and binding "shall" diminishes further the significance of the requirement. The sole mention of the word "Indigenous" in the binding portion of the agreement occurs in Article 7, paragraph 5, which says, in part, "Parties acknowledge that adaptation action . . . should be based on and guided by the best available science and, as appropriate, traditional knowledge, knowledge of indigenous peoples and local knowledge systems."[18] Once again, "should" is used rather than "shall," meaning, "It's up to you, parties."

Justin Trudeau took a leadership role at the COP in a well-publicized effort to insert Indigenous rights into the agreement. It was an early effort to position the Trudeau brand as caring and compassionate. The prime minister said, in a statement before the talks, "I have instructed Canada's chief negotiator for climate change and her team to strongly advocate for the inclusion . . . of language that reflects the importance of respecting the rights of Indigenous peoples."[19]

But there was little hope of this happening and Trudeau must have known it. The powers that be — Big Oil, its corporate allies, and countries representing Big Oil — were stridently opposed to such a development. The US (home of ExxonMobil, Chevron and ConocoPhillips), UK (BP), EU (Royal Dutch Shell, Total, Eni), Norway (Statoil) and Saudi Arabia (Saudi Aramco) all lined up to oppose the inclusion of Indigenous rights. They were concerned about the legal liability that might ensue if greenhouse gas-caused climate change was judged to have violated those Indigenous rights.

Trudeau may have been lucky to fail in his quest to entrench Indigenous rights in the legally binding portion of the text. It would likely preclude him from approving British Columbia's Site C dam, thus endorsing the flooding of the Peace River Valley, a major fishing and hunting site for the Prophet River, West Moberly and other Treaty 8 Tribal Association First Nations. Trudeau didn't adequately consult Indigenous people and recognize their rights before giving the green light on the dam, which will impact climate change, as decomposing materials in the reservoir release methane, a more potent greenhouse gas than carbon dioxide, at least in the short term. In total, a dam emits as much greenhouse gas as would be produced by creating an equivalent amount of electricity by burning natural gas.[20] Nor did Trudeau adequately consult Indigenous peoples before approving Kinder Morgan's Trans Mountain pipeline to the West Coast or Enbridge's Line 3 expansion to Wisconsin.

Paris wasn't the first time that Indigenous peoples had tried to participate in climate change negotiations. But the outcome was always the same: They got lip service.

Big Oil is correct to fear the entrenchment of Indigenous rights in climate change treaties, since it could materially affect business as usual. Yet, legal recognition of such rights seems inevitable as courts in countries with significant Indigenous populations inch forward in their decisions toward such a determination. Canada's belated adoption of the United Nations Declaration on the Rights of Indigenous Peoples may be a major step in that direction. The declaration was adopted in 2007 with 144 states voting in favour, eleven abstaining and four against — Canada, the United States, Australia and New Zealand. The Harper government was concerned that the declaration's requirement for "free, prior and informed consent" for development on Indigenous land amounted to veto power.[21]

The Trudeau government dropped Canada's objector status to the declaration in 2016 and began a period of consultation with First Nations to identify priorities for moving forward on the declaration. Little has come of this process, while Trudeau continued to approve dams, mines and pipelines without obtaining the "free prior and informed consent" promised in the United Nations declaration.

But Parliament is not the only arm of government in Canada. The country's courts have gradually clarified their obligation to take into account Indigenous legal rights.[22] In the landmark Delgamuukw case, the Supreme Court decided in 1997 that "the source of aboriginal title appears to be grounded both in the common law and in the aboriginal perspective on land; the latter includes, but is not limited to, their systems of law."[23] The aboriginal perspective on the occupation of their lands "can be gleaned, in part, but not exclusively, from their traditional laws, because those laws were elements of the practices, customs and traditions of aboriginal peoples." Indigenous law may not be written down, but it is as binding on Indigenous people as legislative acts and judicial decisions are on all Canadians. Emerging from this new understanding is the principle of legal pluralism — distinct but equal laws and legal orders that apply to the same lands and waters.

University of Victoria legal scholar John Borrows says that the "underpinnings of Indigenous law are entwined with the social, historical, political, biological, economic, and spiritual circumstances of each group. They are based on many sources, including sacred teachings, naturalistic observations, positivistic proclamations, deliberative practices, and local and national customs."[24] Leading Canadian Indigenous rights lawyer Louise Mandell makes a similar point: "Indigenous laws and legal orders are rooted in ancestral spirituality. In this spiritual ecology the legal orders include the Creator, supernatural beings, helpers and healers and a culture transmitting orally across generations embodying principles to live by, giving rise to rights and responsibilities," she explained in a 2018 speech.[25] Then-B.C. chief justice Lance Finch told a legal conference in 2012 that the courts — judges and lawyers — coming as they do from a background in British common law or French civil law, are ill-equipped to comprehend the nature of Indigenous law. They have a duty to learn. Reconciliation is not only about making space for Indigenous people, their laws and legal orders, within the known Canadian landscape, but also to enter Indigenous landscapes with an open mind.[26]

Official recognition of Indigenous legal orders, in Canada and worldwide, cannot come soon enough for Indigenous peoples. Climate change not only threatens their traditional way of life but also their rights, as they are no longer able to live according to their traditional legal orders.

The most well-known impacts are those on the Indigenous peoples of the Pacific Islands, who face the prospect of forced migration as rising sea levels swamp their communities. But other examples abound around the world as Indigenous peoples lose the natural basis of their culture, whether it's the Sami of northern Scandinavia (whose reindeer are disappearing) or the peoples of Africa's Kalahari Basin (whose livestock farming is threatened by rising temperatures, dune expansion, increased wind speeds and loss of vegetation). As their cornfields wither, the Kaqchikel Maya of Guatemala are losing both their principal food source and the basis of their culture. The Mi'kmaq of Atlantic Canada and the Gaspé are in a similar quandary as climate change kills their fish and renders unrecognizable the landscapes that form their world and shape their laws and legal order.

Even measures to deal with climate change, such as clean energy projects, can exacerbate the blows already inflicted on Indigenous peoples. Usually they are not consulted, nor are their rights recognized. It turns out that the region of the Sami, with its strong and steady winds and sparse population, is ideal for wind-power projects. A massive US$8.2 billion wind-power development in northern Sweden is blanketing a five-hundred square kilometre area with eight hundred kilometres of roads and more than a thousand giant turbines. The project will remove about a quarter of Sami winter grazing land.[27]

Biofuels — fuels created from plants or various types of waste by-products — is another option for an imagined clean energy future. But in Brazil, Indigenous peoples have been pushed off their lands to make way for vast sugarcane plantations that have turned Brazil into the world's second-largest ethanol producer. The result has been an almost complete dependency on the federal government for food, a vastly reduced life expectancy, and one of the highest suicide rates in the world — what is there to live for? The rivers they use for drinking, bathing and fishing are polluted by chemicals applied to sugarcane crops. A 2009 report by the North American Congress on Latin America says that "most Guarani men now toil" — for pitiful wages in atrocious conditions — "as sugar cane workers on their native land, while their leaders and countless others have been murdered for asserting their rights."[28]

Royal Dutch Shell appeared on the scene in 2010 after signing a US$12 billion joint venture with Cosan, Brazil's largest sugar and ethanol producer. But Cosan was using officially recognized Guarani land to grow its sugarcane.[29] Negative publicity forced Shell to search for "possible solutions." The company "agreed to stop buying sugarcane from land declared as indigenous by the ministry of justice," the activist NGO Survival reported in 2012.[30] While this announcement was spun as good news — and it is, to a certain extent — the fact is that most Guarani traditional land has not been declared Indigenous by the pro-development federal government, and probably never will be.

And then there is hydropower. It is not greenhouse gas-free, because flooded lands release methane; and hydro projects often displace Indigenous people, undermine their culture, remove their food sources and pollute their environment. Here, too, there are examples all over the world and, in Canada, a long history that continues today from Muskrat Falls in Labrador to the Site C dam in British Columbia.

Not every renewable effort goes bad, nor need it. As part of its goal of seeing 30 per cent of electricity in the province generated from renewable sources by 2030, the Alberta government announced in February 2018 an auction for bids to develop wind projects that require Indigenous equity participation. The Kainai (Blood) First Nation in southern Alberta says it is preparing a bid for a project to be built on its land. "We have everything in place. We have the expertise, we've built some capacity to have the financing in place, we're ready to go," Chief Roy Fox told the *Calgary Herald*.[31]

WE CAN LISTEN TO NATURE

Perhaps the greatest challenge to corporate power will come from the movement to imbue Nature with legal rights, to transition from the Western view that nature is mere property to the view that nature consists of rights-bearing entities. The concept flows seamlessly from the special relationship between Indigenous people and their lands. In 2017 the New Zealand parliament passed the *Te Awa Tupua Act*, which gave the Whanganui River and ecosystem a legal standing in its own right, guaranteeing its "health and well-being." The local Maori tribe, consider themselves and the Whanganui, New Zealand's third-longest river, as an indivisible whole.[32]

Other rights-of-nature initiatives have a broader scope. In 2008, under popular socialist president Rafael Correa, Ecuador became the first country to secure the rights of nature in its constitution. Article 71 states that "Nature, or Pacha Mama, where life is reproduced and occurs, has the right to integral respect for its existence and for the maintenance and regeneration of its life cycles, structure, functions and evolutionary processes." Article 72 states that nature has the right to be restored.[33] The new constitution was approved in a referendum in which 69.5 per cent of votes were cast in favour.

In the Ecuadorean conception, anyone can protect nature's rights. In a 2011 case brought in the name of the Vilcabamba River by an American couple that owned property along the river, a provincial court ruled that the rights of the river had been violated by government road construction, which was depositing large quantities of rock and excavation materials in the river.[34]

Ecuador's approach to the rights of nature was soon emulated by Bolivia.

In 2012, Bolivia's Indigenous president, Evo Morales, officially declared the Law of the Rights of Mother Earth. This law established a bill of rights for Mother Earth (which includes ecosystems and human communities), including rights to life, to water, to clean air, to equilibrium and to restoration.[35] At the same time the country was demanding steep carbon emission cuts at UN talks, and was attacked by the UK and the US for so doing.[36] "Recognizing [nature's] rights does not stop all use of nature, but means that our actions must not interfere with the ability of ecosystems and species to thrive," Mari Margil of the US-based Community Environmental Legal Defense Fund and her colleagues explain.[37]

Bolivia hosted the World People's Conference on Climate Change and the Rights of Mother Earth in April 2010. This event brought 35,000 people from 140 countries to Cochabamba to create an alternative to the 2009 United Nations Framework Convention on Climate Change conference in Copenhagen, which failed to come up with a successor to Kyoto. A conference product was the Universal Declaration of Rights of Mother Earth. The declaration "includes the right of Mother Earth and her beings to life and existence, respect, to regenerate its biocapacity and to continue its vital cycles and processes free from human disruptions," explains the Indigenous news service Intercontinental Cry.[38] The declaration has been presented to the UN General Assembly and 800,000 people have signed on to urge the UN to adopt the declaration.

If endowing rivers and other elements of Nature with legal status seems radical, consider the status of corporations. Thanks to a century of judicial decisions by conservative courts, corporations have been given all the rights of persons, including the right to free speech guaranteed in the American *Constitution* and the right to freedom of expression in the *Canadian Charter of Rights and Freedoms*. Corporations are the most powerful force in society, largely because of the legal protections they have squeezed from the courts.

The roots of this perverse situation lie in the decades after the American Civil War. The Fourteenth Amendment to the *Constitution* had been adopted to guarantee all persons "due process and equal protection of the law." It was intended to protect the rights of newly freed slaves, but ended up empowering railroad companies. Real persons had to pay taxes only on the residual value of their property, after mortgages were deducted.

Railroad companies, in contrast, had to pay taxes on the fully assessed value of their holdings. If they could be treated the same as real persons, they could deduct the value of bonds issued by the company. These bonds often exceeded the company's assets, so the railroad company would end up paying little or no tax.

Fourteenth Amendment protection for their corporations was the Holy Grail sought by the railroad barons. They found their Sir Galahad in the guise of an obscure California railroad lawyer who they were able to propel onto the bench of the US Supreme Court. There he proposed a new interpretation of the Fourteenth Amendment, first applied in the famous 1886 case of *Santa Clara County v. Southern Railroad Company*, which allowed the railway company to be treated as a person for tax purposes and thus slashing its taxes.[39] The decision quickly became a precedent used by activist conservative judges to promote the startling proposition that corporations are persons, guaranteed equal protection of law and due process. The century-long campaign culminated in the 2010 Supreme Court decisions in *Citizens United* that applied the *Constitution*'s free speech clause to corporations and opened the floodgate of corporate spending on American elections.[40]

In Canada, the equivalent law is the right to freedom of expression in the *Charter of Rights and Freedoms*. A 2007 Supreme Court of Canada decision "to uphold federal law restricting the advertising of tobacco" seems to be a "welcome endorsement of government's right to control the activities of corporations," writes journalist and author Wade Rowland. The decision wasn't as positive as it seems, he writes. The tobacco companies had argued that "the law limiting advertising infringed on their right to freedom of expression under the *Charter*," and the Court unanimously agreed.

Perhaps one day, ecosystems, too, will have the benefit of this sort of deference, and will have the right to speak for themselves, just as corporations can. The lines are being drawn as nature faces off against corporations. The future of the planet hangs in the balance.

CONCLUSION

Nearly every nation in the world — Canada prominently included — solemnly pledged to start taking serious action to fight global warming at the 2015 Paris conference. Despite these lofty promises, though, carbon dioxide emissions continued their upward trajectory, hitting their highest level ever in 2017, 32.5 gigatonnes of carbon dioxide equivalent.[1] That was followed in early 2018 with the highest concentrations of carbon dioxide in the atmosphere ever recorded at the Mauna Loa Atmospheric Observatory in Hawaii, 410 parts per million.[2] The last week of June and first week of July 2018 witnessed some of the hottest temperatures ever recorded, including the hottest temperature ever reliably recorded in Africa, 51.3 degrees Celsius in Ouargla, Algeria.[3]

These trends are unlikely to change enough to make a significant difference as long as neoliberalism remains the prevailing ideology and Big Oil remains a major player in climate change deliberations.

These developments worry many Canadians, but others seem immune. How can climate change be a serious concern when Canadians continue to buy oversized, gas-guzzling automobiles in record numbers? When first-quarter 2018 sales of electric vehicles were announced, clean energy advocates cheered. The good news was that sales were up 75 per cent over the first quarter of 2017, and accounted for 6,600 vehicles.[4] But that was only 1.5 per cent of total vehicle sales. Light trucks (pick-ups, vans and SUVs), whose emissions have gone up 150 per cent since 1990, accounted for 70 per cent of all vehicle sales in Canada.[5]

A February 2018 online survey of Canadian attitudes to global warming and carbon pricing may explain why Canadians keep buying these vehicles. The poll revealed that a sizeable minority of Canadians still doesn't believe in man-made climate change. The poll, by Abacus Data for the carbon pricing advocacy group Ecofiscal Commission, found that 38 per cent of respondents don't believe there is solid or conclusive evidence that the Earth is warming.[6] Some say there may be some warming, but there is not conclusive evidence for this (27 per cent), while other respondents (11 per cent) say there is little or no evidence for warming. This denial increases to 43 per cent in Ontario, 46 per cent in Saskatchewan and 49 per cent in Alberta and becomes a majority among respondents who voted Conservative in the 2015 federal election (56 per cent).

If respondents accept that the Earth *is* warming, 70 per cent say it is due to human and industrial activity, while 30 per cent say it's caused by natural patterns. That proportion rises to 32 per cent in Ontario, 36 per cent in Saskatchewan, 46 per cent in Alberta and 58 per cent among Conservative voters. Such results indicate that neoliberal denial is alive and well in 2018 Canada.

The survey also contained results that are contra-indicative of neoliberal influence. Respondents were asked how governments should reduce emissions. Forty-four per cent said that government should bring in rules and regulations to reduce emissions in specific sectors and in specific ways. The second option, that governments should provide subsidies to encourage the adoption of low-carbon technologies, was preferred by 35 per cent of respondents. Carbon pricing, the option favoured by most economists, including the Ecofiscal Commission as we've seen, was preferred by just 13 per cent of respondents. Despite wall-to-wall propaganda for carbon pricing, Canadians seem to know it won't work.

That spells trouble for some politicians, opportunity for others. Justin Trudeau and Rachel Notley tied their political horses to Big Oil's cart by executing its so-called grand bargain of pricing and pipelines. They both face the voters in 2019 with lots of smoke and mirrors about the tax, but little real progress on reducing greenhouse gas emissions.

Other politicians — mostly on the right — are taking full advantage of the public's lack of support for carbon pricing. Saskatchewan premier Brad Wall didn't sign Justin Trudeau's Framework Agreement on Clean Growth and Climate Change because it included a carbon tax.

Alberta's opposition leader, Jason Kenney, also opposed the Trudeau-Notley tax plan, skirting the fringe of climate change denial. "We all know climate change exists because . . . the climate's been changing since the beginning of time," he declared at a leadership rally, repeating an early Fraser Institute talking point.[7]

One of Doug Ford's first acts as Ontario's new premier in the summer of 2018 was to cancel the province's cap-and-trade system, which had taken Liberal governments more than a decade to bring to fruition. "The carbon tax is the single-worst tax . . . that anyone could ever have," he said during a leadership debate.[8] He also cut the province's electric vehicle rebate program, slicing deeply into that fragile market.

Meanwhile, federal Environment Minister Catherine McKenna threatened to impose a tax on any province that did not have some carbon pricing plan in effect or in the works. Coast-to-coast carbon pricing was an essential component of Big Oil's national energy strategy.

The carbon pricing battle had been joined, even though carbon pricing is the least effective — and least popular — form of emission reduction. But focusing on pricing, being either for or against, diverted attention away from more direct and effective ways of cutting emissions, Big Oil and its neoliberal allies had created the perfect conditions for business (nearly) as usual to continue.

Progressive Conservative Alberta premier Alison Redford may have talked about a national energy strategy, and her successor Jim Prentice may have been the politician chosen to deliver it, but New Democrat Rachel Notley did the job. Federally, Conservative prime minister Stephen Harper couldn't accept carbon taxes or federal-provincial negotiations and left office without delivering much of anything to Big Oil. It was left to his successor, Justin Trudeau, to bring in the climate plan.

"Getting our oil to new markets while we invest and fund transition towards cleaner energy and more renewables and a greener economy is what Canadians expect and that's what we're going to continue to stand up and do," Trudeau said at the end of a Liberal cabinet retreat in Nanaimo, B.C. in August 2018.[9]

Without caps on oil sands emissions and carbon pricing, there could be no pipelines and without pipelines there could be no national strategy. Ironically it was a social democrat and a liberal, and not a Conservative like Harper, Prentice or Redford, who delivered what was in essence the Business Council of Canada's energy plan.

Most Canadians believe climate change is real and is caused by human and industrial activity. They also believe that direct government regulations and not carbon pricing is the best way to reduce emissions. Any government brave enough to break its neoliberal shackles, stand up to Big Oil and impose tough standards and regulations on recalcitrant industrial polluters will find a receptive audience for its actions.

GLOSSARY

American Petroleum Institute: The largest lobbying and trade association for the American oil and gas industry. It is based in Washington, D.C. with offices in over two dozen state capitals. American Petroleum Institute was a leading promoter of climate change denial.

Brundtland Commission: Formally the *World Commission on Environment and Development*, it was established in 1983 by the United Nations and chaired by Gro Harlem Brundtland, then between terms as prime minister of Norway. Maurice Strong was a prominent member of the commission, which published its influential book, *Our Common Future*, in 1987. This publication established sustainable development as a dominant political force and led to the 1992 *UN Conference on Environment and Development* in Rio.

Business Council of Canada: See *Business Council on National Issues.*

Business Council for Sustainable Development: An organization of forty-eight chief executives of major global corporations established for the 1992 *United Nations Conference on Environment and Development* in Rio. The organization was set up by Swiss industrialist Stephan Schmidheiny on the request of conference director general Maurice Strong to provide the business voice at the conference. It was transformed into the *World Business Council for Sustainable Development* in 1995.

Business Council on National Issues: An organization of the chief executives of 150 major and emerging Canadian corporations that has been influential in shaping Canadian government policies to favour business interests. It was created in 1976 to counter the Trudeau government's wage and price controls. It played an important behind-the-scenes role in shaping Canada's climate policy, especially with the 2007 publication of "Clean Growth: Building a Canadian Environmental Superpower." It changed its name to *Canadian Council of Chief Executives* in 2001 and to *Business Council of Canada* in 2016.

Canada School of Energy and Environment: A school at the University of Calgary established in 2008 to engage in clean energy research. It was headed by Harper government official Bruce Carson, who turned it into an advocacy organization for the Alberta oil sands. When Carson left in 2011 the school reverted to its original mandate.

Canada West Foundation: A Calgary-based think tank established in 1970 by leading business executives to promote western Canadian interests. It helped with the formation of the Reform Party in the 1980s and today primarily develops research to advance the region's key natural resource industries. It played a prominent role in the creation and promotion of a national energy strategy.

Canadian Association of Petroleum Producers: Formed in 1992 by the merger of the *Canadian Petroleum Association* representing large operators and the *Independent Petroleum Association of Canada* representing mid-sized operators, to become a unified voice for the industry. CAPP includes all up-stream operators plus financial services firms, law firms and drilling companies as associates. CAPP promoted voluntary approaches to climate change and lobbied vigorously for oil interests.

Canadian Coalition for Responsible Environmental Solutions: A coalition of twenty-five associations set up by public relations firm National Public Relations in October 2002 to prevent House of Commons ratification of the *Kyoto Protocol*. Its financial backing came largely from the oil and gas industry. It was dissolved in 2011.

Canadian Council of Chief Executives: See *Business Council on National Issues*.
Canadian Petroleum Association: A predecessor to the *Canadian Association of Petroleum Producers*. It was formed in 1952 out of a predecessor organization, the Alberta Oil Operators' Association (1927). Its membership included oil and gas exploration, development, pipeline and transportation companies. It lobbied for their interests in Alberta and Ottawa.

Competitive Enterprise Institute: An American neoliberal think tank founded in 1984 in Washington, D.C. that was and remains a leading voice of climate change denial. Its director of energy and environment led the Donald Trump transition team to select the head of the *Environmental Protection Agency*, choosing Oklahoma attorney general Scott Pruitt, who accepts climate change but casts doubt on the need to fight it.

Conference of Parties: Known as COP, it is the decision-making body responsible for monitoring and reviewing the implementation of the *UN Framework Convention on Climate Change*. It brings together the 197 nations and territories — the Parties — that have signed on to the Framework Convention. The COP has met annually since 1995. It was known as COP1. The Paris Agreement was reached at COP21 in 2015.

Ecofiscal Commission: A Canadian not-for-profit organization formed in 2014 by a group of Canadian economists and embedded in McGill University. Its goals are to promote carbon pricing and the pricing of other environmental risks across the political spectrum.

Energy East pipeline: A plan put forward in 2013 by TransCanada Corporation, to build a CAD$15.7 billion pipeline to ship 1.1 million barrels a day of Alberta bitumen to Saint John, New Brunswick, and from there by tanker to international markets. The company submitted its proposal to the *National Energy Board* in 2015, but ran into significant opposition. With the resurrection of TransCanada's *Keystone XL pipeline* after Donald Trump reauthorized it in 2017, TransCanada withdrew its Energy East proposal.

Energy Policy Institute of Canada: A not-for-profit organization created in 2009 by major oil and gas companies and pipeline operators to be industry's chosen vehicle to promote a national energy strategy and have it adopted by federal and provincial governments. *CAPP* was the primary organizer. The strategy was endorsed by the Harper government in 2011 and was vigorously promoted by Alberta premier Alison Redford. It was dissolved in 2015.

Environmental Defense Fund: A not-for-profit American environmental organization created in 1967 to stop the spraying of DDT in New York State. In the 1980s it became the leading advocate for environmental markets, first for sulphur dioxide that caused acid rain and then for carbon dioxide emissions. With major corporate support, the organization spent US$182 million in 2017.

Environmental Protection Agency: Created by Richard Nixon and authorized by the US Congress in 1970 in the wake of environmental disasters and heightened public concern about environmental pollution. The *EPA* has been a flashpoint of controversy engendered by polluting industry and its surrogates in Congress, think tanks, such as the *Competitive Enterprise Institute* and industry associations such as the *American Petroleum Institute*. EPA foe Scott Pruitt was appointed to head the agency by Donald Trump, resigning in July 2018.

Fraser Institute: A neoliberal think tank with links to the Mont Pèlerin Society. Based in Vancouver, it was created in 1974 to resist the social democratic policies of the B.C. New Democratic government of Dave Barrett and the federal Liberal government of Pierre Trudeau. It produces neoliberal-based research products on a wide range of topics, such as school rankings, economic freedom and deregulation. In 2017 it spent CAD$11.1 million on its activities.

Global Climate Coalition: Created in 1988 by ExxonMobil and the *American Petroleum Institute* and comprising fifty companies in the mining, oil and gas, electricity, automobile, and chemical industries, to counter the work of the *Intergovernmental Panel on Climate Change*. Global Climate Coalition's mission was to convince the US Congress and the media that the idea of human-caused global warming was not true. It opposed the implementation of the *Kyoto Protocol*. It ceased to exist in 2001 after members deserted the organization.

Heartland Institute: A Chicago-based neoliberal think tank with links to the *Mont Pèlerin Society*, founded in 1984. Like many such organizations,

Heartland started out denying the links between smoking and cancer. It is noted for its annual international conference on climate change denial.

Institute of Economic Affairs: The first neoliberal think tank, created in London in 1955 by British businessman Antony Fisher, on the advice of Friedrich Hayek. It is credited with leading to the victory of Margaret Thatcher twenty-five years later. It became the model for hundreds of such think tanks around the world established by wealthy conservative businessmen in the succeeding decades.

Intergovernmental Panel on Climate Change: This body was established by the United Nations Environment Programme and the World Meteorological Organization in 1988 to provide a clear scientific view on the current state of knowledge in climate change and its potential environmental and socio-economic impacts. It produced assessment reports in 1990, 1995, 2001, 2007 and 2014.

International Chamber of Commerce: A Paris-based global organization of national chambers of commerce that promotes business interests at the highest levels, including the *World Trade Organization*, G20 and *UNFCCC*. Issues include free trade, deregulation and privatized dispute-resolution services. It came to play a leading role in climate change negotiations and influenced the *Paris Agreement*.

International Climate Change Partnership: An association of corporations and trade associations that promotes emission-trading schemes at the *UNFCCC* and was one of the powers behind the adoption of the *Kyoto Protocol* at *COP3*. It was created by companies and associations that were party to the 1987 Montreal Protocol to restore the earth's ozone layer.

International Emissions Trading Association: A global organization of oil and gas companies, banks, law firms and consultancies organized by BP in 1999 that promotes carbon markets. Canadian members include ATCO, Enbridge, TransCanada Corporation, Suncor Energy, Capital Power and Ontario Power Generation. The organization was successful in inserting carbon markets into the *Paris Agreement* at *COP21*.

THE BIG STALL

International Energy Agency: A Paris-based organization created in 1974 during the first oil crisis to promote energy security and energy policy co-operation among members of the *OECD*. Today it provides authoritative statistics and analysis on energy issues.

Keystone XL pipeline: A major extension of the original Keystone pipeline proposed by TransCanada Corporation in 2008. XL, which will cost about CAD$10.8 billion, will duplicate the Phase 1 pipeline, but over a shorter route and with larger diameter pipe, giving it the capacity to transport up to 830,000 barrels per day. It will connect TransCanada's terminal at Hardisty, Alberta directly to Steele City, Nebraska, where it would connect with the existing Keystone system. It was rejected by US President Barack Obama in 2016 and resurrected by Donald Trump in 2017. It was approved by the Nebraska Public Service Commission at the end of 2017 but then became subject to an appeal by farmers and landowners along the route.

Kyoto Protocol: An international agreement linked to the UN *Framework Convention on Climate Change*, which commits developed countries to achieve internationally binding emission reduction targets. It provides a variety of emissions trading schemes to help lower emissions. It was adopted in Kyoto, Japan in 1997 and entered into force in 2005. Its first commitment period started in 2008 and ended in 2012.

Line 3 pipeline: A project by Enbridge to replace and expand the entire 1,660-kilometre pipeline that runs from Hardisty, Alberta to Superior, Wisconsin at a cost of CDN$5.3-billion for the Canadian section and US$2.9 billion for the American section. The replacement will expand capacity from about 380,000 to 760,000 barrels per day. The pipeline already has Canadian approval and received Minnesota Public Utilities Commission approval in June 2018.

Mont Pèlerin Society: An organization of free-market economists, business executives and journalists established by Austrian economist and philosopher Friedrich Hayek in 1947 in Mont Pèlerin, Switzerland. Its purpose was to develop theories and strategies to counter the Keynesian orthodoxy that

dominated the intellectual landscape of the postwar period and to disseminate these widely. It could be considered command centre of the neoliberal
revolution.

National Energy Board: The federal regulatory body for international and
interprovincial aspects of major oil, gas and electric utility project proposals. It has been subjected to much criticism over the years, particularly for
regulatory capture (being overly influenced by the companies it's supposed
to regulate), and its inability to consult with Indigenous peoples. It will
be replaced by the Impact Assessment Agency of Canada contained in Bill
C-69, which passed Third reading in June 2018.

National Energy Program: A policy introduced by the Pierre Trudeau government in 1980 to increase the federal share of energy revenues, boost
Canadian ownership of the industry and make Canada self-sufficient as an
oil producer. In the process it transferred economic benefits to Ottawa and
energy users (especially in Ontario and Quebec) at the expense of industry
and the producing provinces. The NEP may have been the most ambitious
effort by a Canadian government to intervene in the economy in recent
history. It was vilified by Big Oil and the oil-producing provinces and ter-
minated by the Brian Mulroney government in 1985.

National Roundtable on the Environment and the Economy: An arm's-length policy advisory body created by the Mulroney government in 1988 to advise the
government how to achieve a sustainable economy. Members were sustainability leaders from across Canada. Its reports, especially on climate change,
were embarrassing to the Harper Conservatives, who killed it in 2013.

Northern Gateway pipeline: A CAD$7.9 billion proposal by Enbridge to build
a twin pipeline from Bruderheim, Alberta to Kitimat, British Columbia.
The eastbound pipeline would import natural gas condensate and the
westbound pipeline would export diluted bitumen to a marine terminal
for transportation to Asian markets. It ran into major opposition from First
Nations and environmental groups. It was approved by the Harper government in 2014, but rejected by the Trudeau government in 2016.

Organization for Economic Cooperation and Development (OECD): A Paris-based international organization composed of thirty-six member nations created in 1961 to support the maintenance and expansion of the market economy. It is a leading advocate for green growth and sustainable development.

Organization of the Petroleum Exporting Countries (OPEC): Formed in 1960 by four countries in the oil-rich Middle East, plus Venezuela. They were joined by other countries in the Middle East and other regions to counter the bargaining power of the Seven Sisters, the world's largest integrated companies, and to stabilize oil markets. Today the fifteen members, with headquarters in Vienna, account for 44 per cent of global oil production and 81.5 per cent of the world's oil reserves, making OPEC a major influencer of global oil prices.

Pan-Canadian Framework on Clean Growth and Climate Change: The plan developed in 2016 by the Canadian government and signed by most provinces to address Canada's commitments to cut greenhouse gases and develop a cleaner economy under the *Paris Agreement* signed at *COP21* of the *UNFCCC*. It requires provinces to implement their own carbon pricing schemes or else have one imposed by the federal government. It follows closely a document published by the *Canadian Council of Chief Executives* in 2007 titled "Clean Growth: Building a Canadian Environmental Superpower."

Paris Agreement: The agreement reached at COP21 in Paris in 2015 as a successor to the *Kyoto Protocol*. Parties agreed to limit climate change to less than two degrees Celsius, but did not adopt mandatory regulations to ensure this would happen. Parties submit intended nationally determined contributions to achieve this goal and aim to undertake rapid reductions in emissions sometime in the future. Emission-trading markets are authorized. The treaty provisions begin in 2020.

Pembina Institute: An environmental organization formed in 1984, first to reduce sulphur emissions in gas plants and later to become an advocate for reducing carbon emissions and promoting clean energy. It is based in Calgary with satellite offices across Canada.

School of Public Policy: Created in 2007 at the University of Calgary with funding from oil and gas lawyer James Palmer to produce research on energy and other topics and train students to work in policy, government and industry. Additional funding came from Imperial Oil. During its first few years, much of the research was directed at supporting the Alberta oil industry.

Smart Prosperity, Smart Prosperity Institute: The Smart Prosperity Leaders Initiative was established in 2016 to track the Trudeau government's clean growth plan. The majority of its members come from business. Most of the others come from non-governmental organizations. The *Sustainable Prosperity Research and Policy Network* was transformed into the *Smart Prosperity Institute* to continue its research and networking activities and to provide research services for the Smart Prosperity Leaders' Initiative.

Sustainable Prosperity Research and Policy Network: This organization was established at the University of Ottawa in 2007 to promote the use of market-based instruments to address environmental problems and to transition into a green economy. Shell Canada was an original sponsor. The organization included a network of researchers at universities across the country. It maintained connections with business organizations, non-governmental organizations and governments, particularly Environment Canada. It was transformed into the *Smart Prosperity Institute* in 2016.

Trans Mountain pipeline: Began operating in 1953 to transmit crude oil from Edmonton to Vancouver and Puget Sound. In 2005, Houston, Texas-based Kinder Morgan bought the company for CAD$6.9 billion, with plans to twin the pipeline and increase capacity to 890,000 barrels per day at a cost of $7.4 billion. The project was opposed by environmentalists, First Nations and the province of B.C. In 2018 the federal government bought the pipeline for CAD$4.5 billion.

United Nations Conference on Environment and Development (Rio Summit, Earth Summit): A major United Nations conference held in Rio de Janeiro in 1992, twenty years after the *United Nations Conference on the Human Environment* in Stockholm. Headed by Canadian Maurice Strong, it was the largest gathering of world leaders in history, with 117 heads of state and representatives of 178 nations attending. Its purpose was to reconcile economic development and environmental protection. Among the documents produced were the Convention on Biological Diversity and the *UN Framework Convention on Climate Change*.

United Nations Conference on the Human Environment: A major United Nations conference held in Stockholm, Sweden in 1972. Headed by Maurice Strong, it was the first international conference on environmental issues to discuss environment and development and led to the formation of the United Nations Environment Programme.

United Nations Framework Convention on Climate Change (UNFCCC): This international environmental treaty was adopted and opened for signature at the *United Nations Conference on Environment and Development* in Rio in June 1992. It entered into force in 1994 when enough countries had ratified it. Its objective was to "stabilize greenhouse gas concentrations in the atmosphere at a level that would prevent dangerous anthropogenic interference with the climate system." It sets non-binding limits on greenhouse gas emissions for individual countries and contains no enforcement mechanisms. Instead, the framework outlines how specific international treaties (called "protocols" or "agreements") may be negotiated to specify further action toward the objective.

World Business Council for Sustainable Development: Created in 1995 by the merger of the *Business Council for Sustainable Development* and the *World Industry Council for the Environment*. Today the organization has about 200 CEO members and is a lead organization in promoting the business viewpoint and in preventing the implementation of mandatory emissions standards at *UNFCCC COPs*.

World Commission on Environment and Development: See *Brundtland Commission*.

ACKNOWLEDGEMENTS

I have relied on the work and guidance of many individuals and organizations in writing *The Big Stall*. I must first acknowledge editor Ted Mumford, who took what he called "ten separate files" and shaped them into this account of Big Oil obstruction over 40 years. Carol Linnitt and Emma Gilchrist at *The Narwhal* (formerly *DeSmog Canada*) commissioned a series of articles on some of the topics that have been incorporated in the book. The associated website *DeSmogBlog* was an invaluable source of comprehensive information on neoliberal think tanks, deniers and Big Oil. My dear friend, Aboriginal and Indigenous rights lawyer Louise Mandell, helped me understand the important role that Indigenous laws and legal orders could play in challenging Big Oil's unrelenting power to shape climate solutions. *The Tyee* has been a constant source of information and inspiration over the years, particularly the work of Andrew Nikiforuk and his ongoing exposes of Big Oil's egregious activities. I published a series of articles called Follow the Money on *Rabble.ca*, which also tracks climate change developments from a progressive viewpoint.

Australian social psychologist Alex Carey's posthumous work, *Taking the Risk Out of Democracy*, first introduced me to the notion of a propaganda-managed democracy and the role neoliberal think tanks play in assisting Big Business to direct society. Philip Mirowski, philosopher of economic thought at Notre Dame University, in his 2013 book, *Never Let a Serious Crisis Go to Waste*, explains how neoliberalism profited from the 2007–2008 financial meltdown after being the main cause of it.

Several scholars helped explain how economists elbowed their way into the climate change policy arena and made carbon markets and taxes the default, if inadequate responses. Environmental economist Raphael Calel provided a critique of the development of emissions trading schemes. Political scientist Jan-Peter Voss of the Technical University of Berlin documented how neoliberal economists created the new reality that carbon emissions could be traded. And international affairs scholar Richard Lane, whose article, "The promiscuous history of carbon trading," also helped clarify how economists created the reality of carbon markets.

Science historians Vladimir Jankovic and Andrew Bowman, in their paper, "After the Green Gold Rush," explained the shift that occurred between 2006–2008 when global warming as threat transitioned to global warming as market opportunity. Ecological economists Clive Spash and Bill Rees provided critiques of mainstream environmental economics and its insistence on growth at any cost. Economist Mark Lee at the Canadian Centre for Policy Alternatives wrote several penetrating critiques of Sustainable Prosperity's laudatory studies of the Gordon Campbell carbon tax. Kevin Taft's book *The Deep State* presented a focused analysis of Big Oil's grip over Alberta and Canada. David Suzuki provided a no-holds-barred take-down of mainstream economics. The David Suzuki Foundation presented ongoing critiques of climate change policy. I didn't cite science historian Naomi Oreskes's ground-breaking work, *Merchants of Doubt*, but it documents the commonalities in the denial efforts of Big Tobacco and Big Oil and other industries that create health, safety and pollution risks.

At James Lorimer and Company, Jim Lorimer encouraged me to write this book, following on from our work in *Harperism*. I also want to thank publisher Carrie Gleason and production editor Sara D'Agostino for their support in bringing this project to fruition. Finally I want to acknowledge my partner Mae Burrows. This is the fifth book she's lived through and she continues to provide good ideas, helpful support, critical reading and delicious, nutritional dinners using vegetables from our garden whenever possible.

ENDNOTES

CHAPTER 1

1. "The history of global oil production," no date. https://sites.google.com/site/globaloilproduction12/1970-s-oil-production.
2. Roy Licklider, "The power of oil," *International Studies Quarterly*, Vol. 32, Iss. 2, 1988, 205–226.
3. Sasha Yusufali, "Petro-Canada," *The Canadian Encyclopedia*, Dec. 16, 2013. https://www.thecanadianencyclopedia.ca/en/article/petro-canada/.
4. "Arab oil embargo," *World Heritage Encyclopedia*, 2018. http://www.self.gutenberg.org/articles/Arab_Oil_Embargo.
5. Gordon Laxer, *After the Sands* (Madeira Park, BC: Douglas and McIntyre, 2015), 39.
6. Sasha Yusufali, *op. cit.*
7. Gordon Laxer, *op. cit*, 136.
8. Sasha Yusufali, *op. cit.*
9. Henry Giniger, "Canada buying unit of Petrofina," *New York Times*, Feb. 3, 1981. http://www.nytimes.com/1981/02/03/business/canada-buying-unit-of-petrofina.html.
10. Jeff Carruthers, "Imperial told to regain oil diverted by US parent," *Globe and Mail*, Feb. 15, 1979, 1.
11. "Alberta and the National Energy Program," CBC Alberta, no date. http://www.cbc.ca/alberta/features/tories40/nep.html.
12. Larry Pratt, "Energy: The roots of national policy," *Studies in Political Economy*, Vol. 7, 1982, 36.
13. Bruce Doern and Glen Toner, *The Politics of Energy: The Development and Implementation of the NEP* (Toronto: Methuen, 1985), 107–108.
14. Larry Pratt, *op. cit.*, 29.
15. Bruce Doern and Glen Toner, *op. cit.*, 206.
16. Gordon Laxer, *op. cit.*, 47.
17. Murray Dobbin, *The Myth of the Good Corporate Citizen* (Toronto: Stoddart, 1998), 166.
18. "Tory discloses deal with premiers on energy debate," *Canadian Press*, Aug. 17, 1984.
19. David Langille, "The Business Council on National Issues and the Canadian state," *Studies in Political Economy*, Vol. 24, 1987, 63.
20. Terence Wills and Jacques Roy, "Oil pact gives firms billion-dollar bonanza," *Montreal Gazette*, Mar. 29, 1985, A1.
21. Bruce Doern and Glen Toner, *op. cit.*, 108.
22. Murray Dobbin, *op. cit.*, 168–75.
23. Paul Chastko, *Developing Alberta's Oil Sands* (Calgary, AB: University of Calgary Press, 2004), 133, quoting a Canadian Petroleum Association report of June 5, 1975.
24. Gordon Laxer, *op. cit.*, 125–26.
25. Peter Lougheed quoted in Allan Tupper, "Peter Lougheed, 1971–1985," in Bradford Rennie (ed.), *Alberta Premiers of the Twentieth Century* (Regina, SK: Canadian Plains Research Center, 2004), 213.
26. Marc Humphries, "North American oil sands: History of development, prospects for the future," Congressional Research Service, Jan. 17, 2008. http://www.oilsandsfactcheck.org/wp-content/uploads/2012/05/CRS-Report-Oil-Sands-History-Future-011708.pdf.
27. "Alberta Gas Trunk Line," Alberta Culture and Tourism, no date. http://history.alberta.ca/energyheritage/gas/transformation/pipelines/alberta-gas-trunk-line.aspx.
28. David Crane, "Ottawa may not care about it but some provinces sure do . . ." *Toronto Star*, Aug. 5, 1989, D2.
29. But it survived the long haul to celebrate its fiftieth anniversary in 2017. It was backed by Sun Oil of Pennsylvania. In 1979 Sun Oil created Suncor Energy by merging all of its

Canadian holdings. Great Canadian Oil Sands was the main asset, along with Sun Oil's conventional oil exploration and production division and the company's refining and retail operations, known as Sunoco. The American parent divested its interest in the mid-1990s.

30. Gordon Laxer, *op. cit.*, 125–126.
31. Paul Chastko, *op. cit.*, 151.
32. Chris Zdeb, "Giant Syncrude project gets the green light," *Edmonton Journal*, Sept. 18, 2015, A2.
33. Robert Page, "Review of Larry Pratt, *The Tar Sands*," *Arctic*, Vol. 29, No. 4, Dec. 1976, 246.
34. Nicholas Kristof, "Canada–U.S. trade tie," *New York Times*, Dec. 11, 1984, 24.
35. Linda McQuaig, *The Quick and the Dead* (Toronto: Penguin Books, 1991), 147.
36. *Ibid.*, 150.
37. Richard Simeon, "Inside the Macdonald Commission," *Studies in Political Economy*, Vol. 22, 1987, 167–171.
38. Eric Uslaner, "Energy policy and free trade in Canada," *Energy Policy*, Vol. 17, No. 4, Aug. 1989, 327–8.
39. Marci McDonald, *Yankee Doodle Dandy* (Toronto: Stoddart, 1995), 221.
40. Quoted in Gordon Laxer, *op. cit.*, 103.
41. It is fallacious to claim, as some do, that the majority of Canadians voted against free trade. The vote wasn't a referendum on a specific question. People vote for political parties for many reasons.
42. Murray Dobbin, *op. cit.*, 47.
43. Dean Henderson, "The four horsemen behind America's oil wars," Global Research, Apr. 26, 2011. https://www.globalresearch.ca/the-four-horsemen-behind-america-s-oil-wars/24507.
44. "Letter from President Roosevelt to King Ibn Saud, April 5, 1945," Crethi Plethi, Nov. 26, 2010. http://www.crethiplethi.com/letter-from-president-roosevelt-to-king-ibn-saud-april-5-1945/usa/2010/.
45. Saeed Kamali Dehghan and Richard Norton-Taylor, "CIA admits role in 1953 Iranian coup," *The Guardian*, Aug. 19, 2013. https://www.theguardian.com/world/2013/aug/19/cia-admits-role-1953-iranian-coup.
46. Peter Newman, "Keeping the frontier alive," *Maclean's*, Jan. 11, 1993, 27.

CHAPTER 2

1. As urged by astroturf expert Richard Berman in his speech to energy industry executives in Colorado Springs in June 2014. See Eric Lipton, "Hard-nosed advice from veteran lobbyist: 'Win ugly or lose pretty,'" *New York Times*, Oct. 30, 2014. https://www.nytimes.com/2014/10/31/us/politics/pr-executives-western-energy-alliance-speech-taped.html.
2. John Stauber, "Review of Bruce Harrison, *Going Green: How to Communicate Your Company's Environmental Commitment*," *EcoAction*, 1994. http://www.eco-action.org/dod/no5/goinggreen.htm.
3. John Stauber and Sheldon Rampton, *Toxic Sludge is Good For You* (Common Courage Press, 2002), 124.
4. Devra Davis, *The Secret History of the War on Cancer* (New York: Basic Books, 2007), 150–51.
5. Kenneth Boulding, "The economics of the coming Spaceship Earth," in Henry Jarrett, ed., *Environmental Quality in a Growing Economy: Essays from the Sixth RFF Forum* (Baltimore: John Hopkins University Press, 1966), 3–14.
6. Nicholas Georgescu-Roegen, *The Entropy Law and the Economic Process* (Cambridge, MA: Harvard University Press, 1971).
7. Robert Nadeau, "Environmental and ecological economics," The Encyclopedia of Earth, Nov. 24, 2011. http://www.eoearth.org/view/article/152604.
8. See the debate between Herman Daly, Stiglitz, Solow and others in *Ecological Economics*, Vol. 22, Iss. 3, Sept. 1997.

9. Herman Daly, *Steady-State Economics* (Washington, DC: Island Press, 1977).
10. E.F. Schumacher, *Small is Beautiful: A Study of Economics as if People Mattered* (London: Abacus Books, 1974), 11.
11. Paul Ehrlich and Anna Ehrlich, *Extinction: The Causes and Consequences of the Disappearance of Species* (New York: Random House, 1981).
12. Markus Peterson *et al.*, "Obscuring ecosystem function with application of the ecosystem services concept," *Conservation Biology*, Vol. 24, No. 1, 2010, 114.
13. Donella Meadows *et al.*, *The Limits to Growth* (New York: Universe Books, 1972).
14. "1972 Stockholm Conference: opening statement," Manitou Foundation, 2017. https://www.mauricestrong.net/index.php/opening statement.
15. "Maurice Strong: Background," Manitou Foundation, 2017. https://www.mauricestrong.net/index.php/list-participants/2-uncategorised/120-strong-short-biography.
16. *Ibid.*
17. Elaine Dewar, *Cloak of Green* (Toronto: Lorimer, 1995), 276.
18. See, for instance, Noam Chomsky, "The Carter administration: Myth and reality," excerpted from *Radical Priorities*, 1981. https://chomsky.info/'priorities01/.
19. Samuel Huntington, "The United States," in Michel Crozier, Samuel Huntington and Joji Watanuki, *The Crisis of Democracy: Report on the Governability of Democracies to the Trilateral Commission* (New York: New York University Press, 1975), 113.
20. Opening discussions with China was also a major concern of the Rockefellers. See Elaine Dewar, *op. cit.*, 278.
21. "Declaration of the United Nations Conference on the Human Environment Report," Stockholm, June 16, 1972. http://www.un-documents.net/unchedec.htm.
22. "World Conservation Strategy," International Union for Conservation of Nature and Natural Resources *et al.*, 1980. https://portals.iucn.org/library/efiles/documents/WCS-004.pdf.
23. Maurice Strong, *Where on Earth are We Going?* (Toronto: Knopf Canada, 2000), 124.
24. *Our Common Future*, The World Commission on Environment and Development (Oxford, UK: Oxford University Press, 1987), 8.
25. Donald Worster, *The Wealth of Nature* (New York: Oxford University Press, 1994), 143.
26. Maarten A. Hajer, *The Politics of Environmental Discourse* (Oxford: Clarendon Press, 1995), 31–32.
27. Friedrich Hayek, *The Road to Serfdom: Text and Documents*, (The definitive edition), ed. Bruce Caldwell (Chicago: University of Chicago Press, 2007), 99.
28. David Harvey, *A Brief History of Neoliberalism* (New York: Oxford University Press, 2005), 2.
29. Timothy Mitchell, "How neoliberalism makes its world," in Philip Mirowski and Dieter Plehwe, eds., *The Road from Mont Pèlerin: The Making of the Neoliberal Thought Collective* (Cambridge, MA: Harvard University Press, 2009), 386.
30. Friedrich Hayek, "The intellectuals and socialism," *University of Chicago Law Review*, Vol. 16, No. 3, 1949, 417.
31. *Ibid.*, 418.
32. *Ibid.*, 419.
33. Timothy Mitchell, *op. cit.*, 387.
34. Patrick Durrant, "Thinking is his business," *The Province*, July 23, 1975, 9.
35. Quoted in Gerald Frost, *Antony Fisher: Champion of Liberty* (London: Profile Books, 2002), 40.
36. *Ibid.*, 71.
37. *Ibid.*, 77.
38. *Ibid.*, 106.
39. *Ibid.*, 103.
40. This section is based partially on Donald Gutstein, "Corporate advocacy: The Fraser Institute," *City Magazine*, Sep. 1978, 32–39.

41. Murray Dobbin, quoted in Brooke Jeffery, *Hard Right Turn* (Toronto: HarperCollins, 1999), 420.
42. Donald Gutstein, *Not a Conspiracy Theory* (Toronto: Key Porter, 2009), 122–23. See also Fraser Institute, "Challenging Perceptions: Twenty-five Years of Influential Ideas," Vancouver, BC, 1999, 4–7.
43. See, for instance, how the Fraser Institute courted the British American Tobacco Co. for specific funding requests, Donald Gutstein, *op. cit.*, 165–67.
44. Gerald Frost, *op. cit.*, 154–60.
45. Hilary Landorf, "Secondary education and sustainable development," in Robert Farrell and George Papagiannis (eds.), *Education for Sustainability* (Oxford, UK: Eolss Publishers, 2009), 181.
46. "The early Keeling curve," Scripps CO2 Program, 2007. https://web.archive.org/web/20070709142518/http://scrippscarbondioxide.ucsd.edu/program_history/early_keeling_curve.html.
47. Quoted in Timothy Mitchell, *Carbon Democracy* (London: Verso, 2013), 6–7.
48. Wallace Immen, "Wayward weather," *Globe and Mail*, Mar. 30, 1981, P19.
49. Philip Hilts, "Global warming trend is forecast by the EPA," *Washington Post*, Oct. 19, 1983, A25.
50. Keetie Sluyterman, "Royal Dutch Shell: Company strategies for dealing with environmental issues," *The Business History Review*, Vol. 84, No. 2, Summer 2010, 203–226.
51. *Ibid.*, 220.
52. "Damian Carrington and Jelmer Mommers, "'Shell knew,'" *The Guardian*, Feb. 28, 2017. https://www.theguardian.com/environment/2017/feb/28/shell-knew-oil-giants-1991-film-warned-climate-change-danger.
53. Brendan DeMelle and Kevin Grandia, "'There is no doubt:' Exxon knew CO2 pollution was a global threat by late 1970s," *The Narwhal*, Apr. 26, 2016. https://thenarwhal.ca/there-no-doubt-exxon-knew-co2-pollution-was-global-threat-late-1970s/.
54. *Ibid.*
55. Neela Banerjee, Lisa Song and David Hasemyer, "Exxon's own research confirmed fossil fuels' role in global warming decades ago," *Inside Climate News*, Sept. 16, 2015. https://insideclimatenews.org/news/15092015/Exxons-own-research-confirmed-fossil-fuels-role-in-global-warming.
56. David Levy and Sandra Rothenberg, "Corporate climate scientists: advocates for science or protectors of status quo?" *The Conversation*, Oct. 2, 2015. https://theconversation.com/corporate-climate-scientists-advocates-for-science-or-protectors-of-status-quo-47991.
57. Neela Banerjee, "Exxon's oil industry peers knew about climate dangers in the 1970s too," *Inside Climate News*, Dec. 22, 2015. https://insideclimatenews.org/news/22122015/exxon-mobil-oil-industry-peers-knew-about-climate-change-dangers-1970s-american-petroleum-institute-api-shell-chevron-texaco.
58. Rob Cunningham, *Smoke and Mirrors: The Canadian Tobacco War* (Ottawa: International Development Research Centre, 1996), 26.
59. Richard Pollay, "Propaganda, puffing and the public interest," *Public Relations Review* Vol. 16, No. 3, 1990, 51.
60. Philip Shabecoff, "Global warming has begun, expert tells Senate," *New York Times*, June 24, 1988, A1; "Statement of Dr. James Hansen, Director, NASA Goddard Institute for Space Studies," presented to United States Senate, June 23, 1988. https://climatechange.procon.org/sourcefiles/1988_Hansen_Senate_Testimony.pdf.
61. Kris M. Wilson, "Communicating climate change through the media," in Stuart Allan, Barbara Adam and Cynthia Carter, eds., *Environmental Risks and the Media* (London: Routledge, 2000), 204.
62. This account relies on Shardul Agrawal, "Context and early origins of the Intergovernmental Panel on Climate Change," *Climatic Change*, Vol. 39, 1998, 605–620.

CHAPTER 3

1. *Our Common Future*, World Commission on Environment and Development (Oxford, UK: Oxford University Press, 1987), 89.
2. Edward Greenspon, "Business leaders contemplate the implications of greenhouse effect," *Globe and Mail*, Feb. 1, 1989.
3. "Environment doomed, politicians told," *Toronto Star*, June 11, 1989, A2.
4. Elaine Dewar, *Cloak of Green* (Toronto: Lorimer, 1995), 293.
5. David Crane, "Conference to address green world challenge," *Toronto Star*, Feb. 7, 1991, C3.
6. Andrew Rowell, *Green Backlash* (New York: Routledge, 1996), 117–118.
7. Matthew Horsman, "Hugh Faulkner follows a new calling," *Financial Post*, Apr. 28, 1989, 18.
8. Kenny Bruno, "The corporate capture of the Earth Summit," *Multinational Monitor*, July, 1992. http://www.multinationalmonitor.org/hyper/issues/1992/07/mm0792_07.html.
9. Philippe Naughton, "Industrialists unveil green business manifesto," *Reuters News*, May 7, 1992.
10. Andre Carothers, "The green machine," *New Internationalist*, 246, Aug. 1993, 14.
11. Ann Reilly Dowd and Suneel Ratan, "How to get things done in Washington," *Fortune*, Aug. 9, 1993, 60.
12. Kevin Watkins, "The foxes take over the hen house," *The Guardian*, July 26, 1992, 27.
13. Geoffrey Palmer, "The Earth Summit: What went wrong at Rio?" *Washington University Law Review*, Vol. 70, No. 4, 1992, 1009. https://openscholarship.wustl.edu/cgi/viewcontent.cgi?article=1867&context=law_lawreview.
14. Stephanie Meakin, "The Rio Summit: Summary of the United Nations Conference on Environment and Development," Ottawa, Library of Parliament, Nov. 1992, 4. https://lop.parl.ca/Content/LOP/ResearchPublicationsArchive/bp1000/bp317-e.asp.
15. David Israelson, "Prepare for disasters in climate, panel says," *Toronto Star*, Aug. 25, 1990, A17.
16. "Article 2: Objective," UN Framework Convention on Climate Change, United Nations, 1992. http://unfccc.int/files/essential_background/background_publications_htmlpdf/application/pdf/conveng.pdf.
17. "The Earth Summit debacle," *The Ecologist*, Vol. 22, No. 4, July/Aug. 1992, 122.
18. Natasha Hassan, "Planting the seeds for a new kind of trade," *Financial Post*, Mar. 18, 1995, 23.
19. Giles Wyburd, "BCSD + WICE = WBCSD," *Business Strategy and the Environment*, Vol. 5, 1996, 48.
20. Nick Cater, "Charter seeks marriage of market and environment," *The Independent*, Apr. 10, 1991, 23.
21. Quoted in David Edward, "Greenwash — Co-opting dissent," *The Ecologist*, Vol. 29, No. 2, May/June 1999, 174.
22. Mary Fagan, "Spending against pollution 'to double,'" *The Independent*, Apr. 13, 1991, 17.
23. Giles Wyburd, *op. cit.*, 49.
24. Philip Mirowski, *Never Let A Serious Crisis Go To Waste* (London: Verso, 2013), 337.
25. David Helvarg, *The War Against the Greens* (San Francisco: Sierra Club Books, 1994), 31.
26. "Total Indoor Environmental Quality," *SourceWatch*, Oct. 15, 2016. https://www.sourcewatch.org/index.php/Total_Indoor_Environmental_Quality.
27. Kenny Bruno, Joshua Karliner and China Brotsky, "Greenhouse gangsters vs. climate justice," Transnational Resource and Action Center, Nov. 1999, 12. http://s3.amazonaws.com/corpwatch.org/downloads/greenhousegangsters.pdf.
28. Kenny Bruno, *op. cit.*
29. Kathy Mulvey and Seth Shulman, "The climate deception dossiers," Union of Concerned Scientists, July 2015, 25. https://www.ucsusa.org/sites/default/files/attach/2015/07/The-Climate-Deception-Dossiers.pdf.
30. "Astroturf," *Sourcewatch*, Oct. 11, 2017. https://www.sourcewatch.org/index.php/Astroturf.

31. Robert Manne, "How vested interests defeated climate science," *The Monthly*, Aug. 2013. https://www.themonthly.com.au/issue/2012/august/1344299325/robert-manne/dark-victory.
32. See Donald Gutstein, *Not a Conspiracy Theory* (Toronto: Key Porter, 2009), chapter 7 and references.
33. Philip Mirowski, *op. cit.*, 338.
34. Studies of neoliberal think tank publications indicate how carefully neoliberal think tanks tracked United Nations and other global developments. For the earlier period see Aaron McCright and Riley Dunlap, "Challenging global warming as a social problem," *Social Problems*, Vol. 47, No. 4, 2000, 499–522. For the middle period, see Jeremiah Bohr, "The 'climatism' cartel," *Environmental Politics*, Vol. 25, No. 5, 2016, 812–830. For the later period, see Constantine Boussalis and Travis Coan, "Text-mining the signals of climate change doubt," *Global Environmental Change*, Vol. 36, 2016, 89–100.
35. McCright and Dunlap, *op. cit.*, Table 1, 508.
36. "Mont Pèlerin Society Directory — 2010." https://www.desmogblog.com/sites/beta.desmogblog.com/files/Mont%20Pelerin%20Society%20Directory%202010.pdf.
37. The story of how markets came to Kyoto and beyond is based on Jonas Meckling, "The globalization of carbon trading," *Global Environmental Politics*, Vol. 11, No. 2, May, 2011.
38. William Stevens, "Scientists say earth's warming could set off wide disruptions," *New York Times*, Sept. 18, 1995, A1.
39. William Stevens, "Industries revisit global warming," *New York Times*, Aug. 5, 1997, A1.
40. Jonas Meckling, *op. cit.*, 33.
41. "Global Climate Coalition," *SourceWatch*, Oct. 11, 2017. https://www.sourcewatch.org/index.php/Global_Climate_Coalition.
42. Philip Mirowski, Jeremy Walker and Antoinette Abboud, "Beyond denial," *Overland*, Issue 210, Autumn 2013, 85.
43. *Ibid.*
44. *Ibid.*, 35.
45. Joby Warrick and Peter Baker, "Clinton details global warming plan," *Washington Post*, Oct. 23, 1997, A1.
46. "The International Climate Change Partnership supports US call for long-term focus on climate change issue," International Climate Change Partnership, Geneva, July 17, 1996. research.greenpeaceusa.org/?a=download&d=2984.

CHAPTER 4

1. Ross Howard, "Call to link arms replaces battle of words at environment meeting," *Globe and Mail*, Dec. 28, 1989, A8.
2. See Kaija Belfry Munroe, *Business in a Changing Climate* (Toronto: University of Toronto Press, 2016), 53; George Hoberg and Kathryn Harrison, "It's not easy being green," *Canadian Public Policy*, Vol. 20, No. 2, June 1994: 125; Anne McIlroy, "Environmental plan 'weakened,'" *Ottawa Citizen*, Mar. 28, 1990, A4.
3. "Towards a sustainable and competitive future," Business Council on National Issues, Task Force on Sustainable Development, May 1992. http://thebusinesscouncil.ca/publications/towards-a-sustainable-and-competitive-future/.
4. Noranda Inc., "Announcement" *Financial Post*, June 28, 1995, 9.
5. "David Manning," Wilson Center, no date. https://www.wilsoncenter.org/person/david-manning.
6. "Climate change: A strategy for voluntary business action," Business Council on National Issues, Task Force on the Environment, Nov. 1994. http://thebusinesscouncil.ca/publications/climate-change-a-strategy-for-voluntary-business-action/.
7. Norm Ovenden, "Energy producers close to a deal on cutting pollution," *Edmonton Journal*, Nov. 22, 1994, B5.
8. Douglas Macdonald, *Business and Environmental Politics in Canada* (Peterborough, ON: Broadview Press, 2007), 134–5.

9. Eric Beauchesne, "Big business to consent to voluntary limits on greenhouse gases," *Montreal Gazette*, Sep. 28, 1995, D3.
10. Matthew Bramley, "The case for Kyoto: The failure of voluntary corporate action," Pembina Institute, Oct. 2002, 2. http://www.pembina.org/reports/VCR_publication_101702.pdf.
11. *Ibid.*, 24.
12. Sheldon Alberts, "Resist pressure on gas, Ottawa told," *Calgary Herald*, Oct. 10, 1997, A2.
13. John Geddes, "Global warming deal penalizes Canada: BCNI," *Financial Post*, July 4, 1997, 8.
14. Hugh Winsor, "The power game: d'Aquino fears the sky will fall," *Globe and Mail*, Nov. 3, 1997, A4.
15. John Geddes, "It's not easy being green," *Financial Post*, July 5, 1997, 14.
16. Peter Newman, "A man of influence," *Canadian Business*, Sept. 29, 2009, 67.
17. Brent Jang, "Oil firms split on global warming," *Globe and Mail*, Dec. 5, 1997, A22.
18. Brent Jang, "Ottawa's greenhouse gas target draws fire," *Globe and Mail*, Dec. 2, 1997, B7.
19. Douglas Macdonald, *op. cit.*, 165.
20. "How do you spell relief?" *Oilweek*, Vol. 47, Iss. 17, Apr. 1996, 5.
21. Alan Boras and Sheldon Alberts, "Oil producers laud tax freeze," *Calgary Herald*, Mar. 7, 1996, A12.
22. Alan Boras, "Plea made to oilpatch," *Calgary Herald*, Mar. 14, 1996, A2.
23. Quoted in Paul Chastko, *Developing Alberta's Oil Sands* (Calgary, AB: University of Calgary Press, 2004), 219.
24. See Marci McDonald, "The man behind Stephen Harper," *The Walrus*, Oct. 12, 2004. https://thewalrus.ca/the-man-behind-stephen-harper/. See also Donald Gutstein, *Not A Conspiracy Theory* (Toronto: Key Porter, 2009), 150–59.
25. "2001 Annual Report," Fraser Institute, Vancouver, 7–8.
26. Lorne Gunter, "An evening with premiers past, present and future," *National Post*, Oct. 16, 2004, A13.
27. Brent Jang, "Oilpatch looks south," *Globe and Mail*, Dec. 9, 1995, B1.
28. "Paul Precht," Oil Sands Oral History Project, Petroleum History Society, Jan 29, 2013, 11. https://www.glenbow.org/collections/search/findingAids/archhtm/extras/oilsands/Precht_Paul.pdf.
29. Dan Woynillowicz, Chris Severson-Baker and Marlo Raynolds, "Oil Sands Fever," Pembina Institute, Nov. 2005, 3. https://www.pembina.org/reports/OilSands72.pdf.
30. Gretchen Ziegler, "Oilsands: Prosperity assured," *Commerce News*, Vol. 18, Iss. 3, Mar. 1, 1996, 1.
31. Paul Chastko, *op. cit.*, 220.
32. Alan Boras, "Oilsands boost worth $5 billion," *Calgary Herald*, June 4, 1996, A1.
33. Brent Jang, "The oilsands' $20.6 billion scorecard," *Globe and Mail*, Dec. 11, 1997, B4.
34. Peter Newman, "The Yanks shouldn't take us for granted," *National Post*, Apr. 16, 2005, A18.
35. Claudia Cattaneo, "Canada's top oil entrepreneur joins Wayne Gretzky, Brian Mulroney and Prem Watsa as winner of prestigious international award." *National Post*, Nov. 18, 2013. http://business.financialpost.com/commodities/energy/canadas-top-oil-entrepreneur-joins-wayne-gretzky-brian-mulroney-and-prem-watsa-as-winner-of-prestigious-international-award.
36. Max Fawcett, "The artist of the deal," *Report on Business Magazine*, Nov. 12, 2017. https://www.theglobeandmail.com/report-on-business/rob-magazine/the-elusive-billionaire-who-wants-to-make-the-oil-sands-greener/article32955602/.
37. Stephen Ewart and Deborah Yedlin, "$1.6 billion deal," *Calgary Herald*, Aug. 6, 1999, C1.
38. Claudia Cattaneo, "Boom times return to the oilpatch capital," *National Post*, Aug. 18, 1999, C10.

CHAPTER 5

1. Bertrand Marotte, "Business doesn't buy curbs," *Hamilton Spectator*, Dec. 13, 1997, D1.
2. Thomas d'Aquino, "Kyoto and the Canadian Challenge," A presentation to the National Forum on Climate Change and the National Roundtable on the Environment and the Economy, Business Council on National Issues, Ottawa, Feb. 17, 1998. http://thebusinesscouncil.ca/wp-content/uploads/archives/presentations_1998_02_17.pdf.
3. "Bravo to Shell on greenhouse-gas initiative," *Toronto Star*, May 4, 1999. 1.
4. Wendy Warburton, "Carney denies there's conflict over reason for dropping gas prices," *Ottawa Citizen*, Mar. 15, 1986, A3.
5. Martin Mittelstaedt, "Companies seek to delay emissions reduction," *Globe and Mail*, Jan 24, 2000, A3.
6. Alan Toulin, "Oilpatch will be hammered by Kyoto Protocol, study says," *National Post*, Dec. 4, 2001, FP5.
7. Dennis Bueckert, "Chrétien reiterates commitment to ratify Kyoto climate treaty next year," *Canadian Press*, Nov. 2, 2001.
8. Kate Jaimet, "Feds say Kyoto to cost 3 cents a barrel," *National Post*, Oct. 23, 2002, FP1.
9. Douglas Macdonald, *Business and Environmental Politics in Canada* (Peterborough, ON: Broadview Press, 2007), 1.
10. Carol Howes, "Canada won't sign Kyoto: Peterson," *National Post*, Mar. 13, 2002, FP5; Brent Jang, "Imperial Oil executives mince no words on Kyoto or anything else," *Globe and Mail*, Mar. 13, 2002, B9.
11. Alan Toulin, "Alberta offers Kyoto alternative," *National Post*, Mar. 28, 2002, FP1.
12. Eva Ferguson, "Klein cools off on Kyoto challenge," *Calgary Herald*, Apr. 19, 2002, A1.
13. Canadian Coalition for Responsible Environmental Solutions, "Coalition formed to advance 'Made in Canada' strategy on climate change," *Canada Newswire*, Sept. 26, 2002; Steven Chase, "Business groups opposed to Kyoto, *Globe and Mail*, Sept. 27, 2002, B3; ; "Canadian Coalition for Responsible Environmental Solutions," *SourceWatch*, Sept. 11, 2008. https://www.sourcewatch.org/index.php/Canadian_Coalition_for_Responsible_Environmental_Solutions.
14. Paul Haavardsrud, "Buckee criticizes oil lobby on Kyoto," *National Post*, Feb. 1, 2003, FP3.
15. Steven Chase, *op. cit.*
16. Hugh Winsor, "Oil patch candour needed on ad campaign," *Globe and Mail*, Nov. 13, 2002, A4.
17. Steven Chase, "Liberal MPs want details on Kyoto," *Globe and Mail*, Oct. 10, 2002, A7.
18. "CCRES calls for consultation on climate change," *Canada Newswire*, Nov. 5, 2002.
19. Patrick Brethour, "Support for Kyoto plunges," *Globe and Mail*, Nov. 2, 2002, A1.
20. Andrew Chung, "Alberta's Kyoto poll 'fraudulent': Anderson," *Toronto Star*, Nov. 5, 2002, A7.
21. Darrell Bricker, "The Kyoto numbers," *Globe and Mail*, Nov. 7, 2002, A24; see also JillMahoney, "Ipsos-Reid lashes out at critics of Kyoto poll," *Globe and Mail*, Nov. 7, 2002, A9.
22. Hugh Winsor, "Questions lurk beneath surface of ads on Kyoto," *Globe and Mail*, Nov. 8, 2002, A4.
23. "The Advancement of Sound Science Coalition," *SourceWatch*, July 1, 2007. http://www.sourcewatch.org/index.php?title=The_Advancement_of_Sound_Science_Coalition.
24. See Donald Gutstein, *Not A Conspiracy Theory* (Toronto: Key Porter, 2009), 235–38.
25. Frank Graves, "Kyoto support stable," *Globe and Mail*, Nov. 6, 2002, A16.
26. Kaija Belfry Munroe, *Business in a Changing Climate* (Toronto: University of Toronto Press, 2016), 61.
27. Paul Willcocks, "Think political donations are benign? You must be a politician," *The Tyee*, Aug. 25, 2014. https://thetyee.ca/Opinion/2014/08/25/Political-Donations-Not-Benign/
28. Dennis Bueckert, "Government waters down targets for emissions cuts by large polluters," *Canadian Press*, Mar. 30, 2005.

29. Douglas Macdonald, *op. cit.*, 3.
30. Laura Jones, ed., *Global Warming: The Science and the Politics* (Vancouver: Fraser Institute, 1997), 17.
31. Laura Jones, "Global warming makes me hot," *Vancouver Sun*, Oct. 28, 1997, A19.
32. Margaret Munro, "Making sense of the global warming issue," *Vancouver Sun*, Nov. 28, 1997, A21.
33. Michael Sanera and Jane Shaw, *Facts, Not Fear* (Vancouver: Fraser Institute, 1999).
34. Quoted by ExxonSecrets, "Factsheet: Reason Foundation," no date. http://www.exxonsecrets.org/html/orgfactsheet.php?id=63.
35. Fraser Institute, "The Politics, Science and Economics of Kyoto," *Fraser Forum*, Jan. 2003.
36. Bill Graveland, "Former Ontario premier makes headlines in first public speaking appearance," *Canadian Press*, Nov. 6, 2002.
37. Ric Dolphin, "Fraser Institute welcome's Klein's Kyoto revolution," *Calgary Herald*, Nov. 16, 2002, A3.
38. Kenneth Green, "Alberta climate policy solid start," *Prince Albert Daily Herald*, Nov. 18, 2002, 4.
39. Kevin Grandia, "What's an IPCC Expert Reviewer?" *DeSmog Canada*, Jan. 11, 2007. https://www.desmogblog.com/whats-an-ipcc-expert-reviewer.
40. "Poll finds Ontario's largest corporations want ratification of Kyoto delayed until detailed plan available and accepted by the provinces," *Canada Newswire*, Nov. 26, 2002, 1.
41. "Science championing Kyoto anything but exact," *The Province*, Dec. 3, 2002, A14.
42. Kenneth Green, "Coal for Christmas courtesy of Kyoto," *National Post*, Dec, 16, 2002, A18.
43. Kenneth Green, Tim Ball and Steven Schroeder, "The science isn't settled," *Fraser Institute, Public Policy Sources*, No. 80, June 2004. https://www.fraserinstitute.org/studies/science-isnt-settled-limitations-global-climate-models.
44. Environmental Literacy Council, "Key supporters," July 3, 2007. https://web.archive.org/web/20070711045353/http://www.enviroliteracy.org/article.php/701.html.
45. Ian Sample, "Scientists offered cash to dispute climate study," *The Guardian*, Feb. 2, 2007. http://www.guardian.co.uk/environment/2007/feb/02/frontpagenews.climatechange.
46. Ross McKitrick *et al.*, "Independent summary for policymakers: IPCC fourth assessment report," Fraser Institute, Feb. 5, 2007, 52. https://www.fraserinstitute.org/studies/independent-summary-for-policymakers.
47. The institute published its membership lists in the early years. See Cliff Stainsby and John Malcolmson, "The Fraser Institute, the government and a corporate free lunch," Solidarity Coalition, Oct. 1983.
48. Donald Gutstein, "Follow the money, Part 5 – The Tobacco Papers revisited," *Rabble*, Apr. 24, 2014. http://www.rabble.ca/blogs/bloggers/donald-gutstein/2014/04/follow-money-part-5-tobacco-papers-revisited.
49. Sherry Stein, Letter to Martin Broughton regarding research program in emulation of the social affairs unit, Jan. 28, 2000. https://www.industrydocumentslibrary.ucsf.edu/tobacco/docs/#id=sjxl0204.
50. Fraser Institute, "Annual Report 1999." https://web.archive.org/web/20020517045057/http://oldfraser.lexi.net:80/about_us/annual_reports/1999/index.html.
51. *Ibid.*
52. Philip Morris International, "External Affairs," Presentation to PMI Corporate Affairs, Aug. 1, 2000. https://www.industrydocumentslibrary.ucsf.edu/tobacco/docs/#id=ymfd0061.
53. Terry Pechacek and Stephen Babb, "Environmental tobacco smoke and tobacco related mortality in a prospective study of Californians, 1960–98," *British Medical Journal*, Vol. 326, No. 7398, May 17, 2003.

54. Sherry Stein, *op. cit.*
55. Laura Jones, Letter to Adrian Payne regarding copy of book published, Fraser Institute, May 18, 2000. https://www.industrydocumentslibrary.ucsf.edu/tobacco/docs/#id=zjxl0204.
56. Michael Walker, Letter to Adrian Payne regarding centre for studies in risk and regulation, British American Tobacco, June 19, 2000. https://www.industrydocumentslibrary.ucsf.edu/tobacco/docs/#id=tjxl0204.
57. *Ibid.*
58. Sherry Stein, Letter to Adrian Payne regarding the media coverage entitled "The History of Tobacco Regulation," Fraser Institute, Sept. 12, 2000. https://www.industrydocumentslibrary.ucsf.edu/tobacco/docs/#id=hhxl0204.

CHAPTER 6

1. Don Drummond, Nancy Olewiler and Christopher Ragan, "Carbon price vs. regulation: The better choice is clear," *Globe and Mail*, Oct. 5, 2016, B4.
2. "Arthur Cecil Pigou," Library of Economics and Liberty, 2002. http://www.econlib.org/library/Enc1/bios/Pigou.html.
3. Janet Milne and Mikael Andersen, "Introduction to environmental taxation concepts," in Janet Milne and Mikael Andersen (eds.), *Handbook of Research on Environmental Taxation* (Cheltenham, UK: Edward Elgar Publishing, 2012), 18.
4. "Arthur Cecil Pigou," *Institute for New Economic Thinking*, no date. http://www.hetwebsite.net/het/profiles/pigou.htm.
5. Jennifer Brown, "Environmental economics," *Encyclopedia Britannica*, Feb. 5, 2016. https://www.britannica.com/topic/environmental-economics.
6. *Ibid.*
7. Thomas Crocker, "The structure of atmospheric pollution control systems," in Harold Wolozin, ed., *The Economics of Air Pollution* (New York: WW Norton, 1966), 61–86.
8. John Dales, *Pollution, Property and Prices* (Toronto: University of Toronto Press, 1968)
9. *Ibid.*, 111.
10. Richard Lane, "The promiscuous history of market efficiency," *Environmental Politics*, Vol. 21, No. 4, July 2012, 585.
11. David Vogel, *Fluctuating Fortunes* (New York: Basic Books, 1989), 61.
12. *Ibid.*, 69.
13. R.J. Powell and L.M. Wharton, "Development of the Canadian Clean Air Act," *Journal of the Air Pollution Control Association*, Vol. 32, No. 1, 1982: 62–65.
14. David Vogel, *op. cit.*, 133.
15. Barry Burke, "Antonio Gramsci, schooling and education," *The Encyclopedia of Informal Education*, 2005. http://infed.org/mobi/antonio-gramsci-schooling-and-education/.
16. *Ibid.*
17. Richard Lane, *op. cit.*, 591.
18. Richard Andrews. "Economics and environmental decisions, past and present," in V. Kerry Smith, ed., *Environmental Policy Under Reagan's Executive Order* (Chapel Hill NC: University of North Carolina Press, 1984), 58.
19. Donald Mackenzie, Material Markets (Oxford, UK: Oxford University Press, 2009), 139.
20. Steven Marcus, "Bubble policy: pros and cons," *New York Times*, June 30, 1983. http://www.nytimes.com/1983/06/30/business/technology-bubble-policy-pros-and-cons.html; Richard Dowd, "EPA's air quality bubble revisited," *Environmental Science and Technology*, Vol. 18, No. 8, 1984: 249A .
21. Raphael Calel, "Carbon markets: a historical overview," *WIREs Climate Change*, Vol. 4, 2013, 109.
22. Jan-Peter Voss, "Innovation processes in governance," *Science and Public Policy*, Vol. 34, No. 5, June 2007, 334.
23. Allan Bullock and Stephen Trombley, eds., *The New Fontana Dictionary of Modern Thought*, 3rd ed. (London: HarperCollins, 1999), 708.

24. *Ibid.*, 251.
25. Richard Lane, *op. cit.*, 593.
26. Philip Shabecoff, "There must be something besides rules, rules, and more rules," *New York Times*, Aug. 21, 1977, 148.
27. David Harvey, *A Brief History of Neoliberalism* (Oxford, UK: Oxford University Press, 2005), 23–25.
28. "Jimmy Carter: Executive Order 12044, Improving government regulations," The American Presidency Project, Mar. 23, 1978. http://www.presidency.ucsb.edu/ws/?pid=30539.
29. "Jimmy Carter: Memorandum from the President on alternative approaches to regulation," The American Presidency Project, June 13, 1980. http://www.presidency.ucsb.edu/ws/?pid=44568.
30. *Ibid.*
31. Charles Schultze, *The Private Use of Public Interest* (Washington DC: Brookings Institution, 1977), 2.
32. Brian O'Keefe, "The red tape conundrum," *Fortune*, Oct. 20, 2016. http://fortune.com/red-tape-business-regulations/.
33. "Our nation's air: Status and trends through 2015," Environmental Protection Agency, 2016. https://gispub.epa.gov/air/trendsreport/2016/.
34. Richard Conniff, "The political history of cap and trade," *Smithsonian Magazine*, Apr. 2009. http://www.smithsonianmag.com/science-nature/the-political-history-of-cap-and-trade-34711212/.
35. Eric Pooley, *The Climate War* (New York: Hyperion, 2010), 65.
36. Fred Krupp, "New environmentalism factors in economic needs," *Wall Street Journal*, Nov. 20, 1986, 34.
37. Eric Pooley, *op. cit.*, 66. In its ultimately unsuccessful campaign between 2005 and 2009 to persuade the US Congress to pass cap-and-trade legislation, EDF received more than $40 million from Julian Robertson, founder of the Tiger Management hedge fund. See Matthew Nisbet, "Environmental Defense Fund mega-donor is biggest individual contributor to Romney super PAC. *Big Think*, Feb. 22, 2012. http://bigthink.com/age-of-engagement/environmental-defense-fund-mega-donor-is-biggest-individual-contributor-to-romney-super-pac.
38. Jan-Peter Voss, "Realizing instruments," in Jan-Peter Voss *et al.*, *Knowing Governance* (London: Palgrave Macmillan, 2016), 137.
39. *Ibid.*, 138.
40. Claude Engle and Hawley Truax, "The carrot or the stick?" *Environmental Action*, Vol. 21, Issue 6, May/June 1990.
41. *Ibid.*, 138; A. Denny Ellerman *et al.*, *Markets for Clean Air* (Cambridge, UK: Cambridge University Press, 2000), 104.
42. Morgan Robertson, "Jimmy Carter and the archeology of market environmentalism," Part I, Wetlandia, July 30, 2013. http://wetlandia.blogspot.com/.
43. Denny Ellerman *et al.*, *op. cit.*
44. *Ibid.*, 315.
45. *Ibid.*, 3.
46. "Our nation's air," *op. cit.*
47. "The economists' statement on climate change," *Redefining Progress*, Mar. 29, 1997. http://rprogress.org/publications/1997/econstatement.htm.
48. Peter Passell, "Economic scene," *New York Times*, Feb. 13, 1997, D2.
49. Barry Brook and Tim Kelly, "Greenhouse tax versus greenhouse cap and trade: The debate we never had," Joint submission to the Senate Standing Committee on Economics," Mar. 25, 2009. http://www.aph.gov.au/~/media/wopapub/senate/committee/economics_ctte/completed_inquiries/2008_10/cprs_09/submissions/sub33_pdf.ashx.
50. Brian Fisher *et al.*, "An economic assessment of policy instruments for combatting

climate change," in *Climate Change 1995, Economic and Social Dimensions of Climate Change, Contribution of Working Group III to the Second Assessment Report of the IPCC* (Cambridge: Cambridge University Press, 1996), 401.

51. Jan-Peter Voss, *op. cit.*, 134.
52. Jane Andrew, Mary Kaidonis and Brian Andrew, "Carbon tax: Challenging neoliberal solutions to climate change," *Critical Perspectives in Accounting*, Vol. 21, 2010: 611–618.
53. Raphael Calel, *op. cit.*, 113.
54. See, for instance, "Guide to carbon trading crime," *Interpol Environmental Crime Programme*, June 2013. http://www.interpol.int/content/download/20122/181158/version/3/file/Guide_to_Carbon_Trading_Crime.pdf.
55. Raphael Calel, *op. cit.*, 107.
56. John Hilsenrath, *op. cit.*
57. *Ibid.*
58. Raphael Calel, *op. cit.*, 108.
59. Greg Mankiw, "Pigovian questions," *Greg Mankiw's Blog*, Dec. 9, 2006. http://gregmankiw.blogspot.ca/2006/12/pigovian-questions.html.
60. John Schwartz, "'A conservative climate solution,' Republican group calls for carbon tax," *New York Times*, Feb. 7, 2017. https://www.nytimes.com/2017/02/07/science/a-conservative-climate-solution-republican-group-calls-for-carbon-tax.html.
61. Barkley Rosser, "Why is the Pigou Club so out of synch with the Paris climate negotiators?" *Econospeak*, Dec. 7, 2015. http://econospeak.blogspot.ca/2015/12/why-is-pigou-club-so-out-of-synch-with.html.
62. Jeffrey Simpson, "Why BC's carbon model is the way to go," *Globe and Mail*, Mar. 4, 2016, A11.
63. Cited in Samuel Randalls, "Optimal climate change," *Osiris*, Vol. 26, 2011, 240.

CHAPTER 7

1. "Climate change 2007: Synthesis report," Intergovernmental Panel on Climate Change, 2008, 2. https://www.ipcc.ch/pdf/assessment-report/ar4/syr/ar4_syr_spm.pdf.
2. Benjamin Stephan, "The power in carbon," Universitat Hamburg, Working Paper Series, No. 4, Jan. 2011. https://www.wiso.uni-hamburg.de/fachbereich-sowi/professuren/engels/archiv/working-papers/wps-no04.pdf.
3. Jonas Meckling, "The globalization of carbon trading," *Global Environmental Politics*, Vol. 11, No. 2, May, 2011, 36.
4. Hiromi Sasamoto, "Industrial leaders open talks on warming," *The Daily Yomiuri*, Dec. 4, 1997, 3.
5. Terry Macalister and Paul Brown, "Earth summit agenda 'hijacked,'" *The Guardian*, Aug. 9, 2002, 2.
6. Jeremiah Bohr, "The 'climatism' cartel," *Environmental Politics*, Vol. 25, No. 5, 2016, 813.
7. Dennis Shanahan, "New Asia-Pacific climate plan," *The Australian*, July 27, 2005, 1.
8. Geoffrey Lean, "Brown counters Bush global warming snub with own global study," *The Independent*, July 31, 2005, 16.
9. "Stern Review on the Economics of Climate Change: Final Report," Great Britain, HM Treasury, 2006, vi. http://webarchive.nationalarchives.gov.uk/+/http://www.hm-treasury.gov.uk/independent_reviews/stern_review_economics_climate_change/stern_review_report.cfm.
10. H. Sterling Burnett, "Climate alarmist quits British government," Heartland Institute, Mar. 1, 2007. https://www.heartland.org/news-opinion/news/climate-alarmist-quits-british-government.
11. Vladimir Jankovic and Andrew Bowman, "After the green gold rush," *Economy and Society*, Vol. 43, No. 2, May 2014, 240.
12. "Stern Review," *op. cit.*, viii.
13. Jeremy Warner, ""It makes good business sense to cut emissions," *The Independent*, Oct. 28, 2006, 54.

14. "Climate change: Everyone's business," *Confederation of British Industry*, Nov. 2007. http://www.cbi.org.uk/pdf/climatereport2007full.pdf.
15. Jankovic and Bowman, *op. cit.*, 243.
16. Paul Sisco, "NASA scientist says global warming danger amounts to 'planetary emergency,'" *Voice of America*, Dec. 9, 2009. https://www.voanews.com/a/a-13-2008-10-08-voa27-66732472/561441.html.
17. Andrew Hoffman and John Woody, *Climate Change: What's Your Business Strategy?* (Cambridge, MA: Harvard Business Press, 2008), 1.
18. John Llewellyn, "The business of climate change: Challenges and opportunities," Lehman Brothers, Feb. 2007. http://www.lehman.com/press/pdf_2007/TheBusinessOfClimateChange.pdf.
19. "Secretary-General in message to Investor Summit for Climate Risk,, says financial leaders see chance to usher in age of green economics, sustainable development," United Nations, Feb. 14, 2008. http://www.un.org/press/en/2008/sgsm11417.doc.htm.
20. Green New Deal Group, "A green new deal," *New Economics Foundation*, July 20, 2008. http://neweconomics.org/2008/07/green-new-deal/.
21. Quoted in Edouard Stenger, "A global green new deal: the United Nations Environment Programme initiative," Oct. 27, 2008. http://www.edouardstenger.com/2008/10/27/a-global-green-new-deal-unep-iniative/.
22. Edward Barbier, "How is the Global Green New Deal going?" *Nature*, Vol. 464, Apr. 8, 2010, 832.
23. "A climate for recovery," HSBC Global Research, Feb. 25, 2009. http://globaldashboard.org/wp-content/uploads/2009/HSBC_Green_New_Deal.pdf.
24. "The Copenhagen communiqué on climate change," The Prince of Wales's Corporate Leaders Group on Climate Change, Cambridge, UK, 2009. https://www.cisl.cam.ac.uk/publications/publication-pdfs/the-copenhagen-communique.pdf.
25. Kenneth Arrow *et al.*, "Is there a role for benefit-cost analysis in environmental health and safety regulation?" *Science*, Vol. 272, Iss. 5259, Apr. 12, 1996, 221–222.
26. Allan Bullock and Stephen Trombley, eds., *The New Fontana Dictionary of Modern Thought*, 3rd ed. (London: HarperCollins, 1999), 178.
27. Paul Baer and Clive Spash, "Cost-benefit analysis of climate change: Stern revisited," CSIRO Working Paper Series 2008-07, May 2008. https://ideas.repec.org/p/cse/wpaper/2008-07.html.
28. Samuel Randall, "Optimal climate change," *Osiris*, Vol. 26, No. 1, 2011, 230.
29. Baer and Spash, *op. cit.*
30. "Winning the battle against global climate change," Council of the European Union, European Commission press release, Brussels, Feb. 9, 2005. http://europa.eu/rapid/press-release_MEMO-05-42_en.htm.
31. Clive Spash, "The economics of avoiding action on climate change," *Adbusters*, 75, Jan./Feb. 2008. http://www.clivespash.org/wp-content/uploads/2015/04/2008_Spash_Adbusters_finalcopy.pdf.
32. David Leonhardt, "Amid the ivy, a battle about the climate," *New York Times*, Feb. 21, 2007, C1.
33. *Ibid.*
34. "Is it worth it?" *The Economist*, Dec. 3, 2009. http://www.economist.com/node/14994731.
35. Leonhardt, *op. cit.*
36. Charles Komanoff, "Is the rift between Nordhaus and Stern evaporating with rising temperatures?" Carbon Tax Center, Aug. 21, 2014. https://www.carbontax.org/blog/2014/08/21/is-the-rift-between-nordhaus-and-stern-evaporating-with-rising-temperatures/.
37. Joe Romm, "Once-cautious climate economist reverses course, issues warning against the cost of inaction," *Think Progress*, Feb. 4, 2017. https://thinkprogress.org/once-cautious-climate-economist-warns-against-the-cost-of-trump-era-inaction-c19ff36ff964/.

38. Andrea Saltelli and Beatrice D'Hombres, "Sensitivity analysis didn't help. A practitioner's critique of the Stern review," *Global Environmental Change*, Vol. 20, 2010, 298.

CHAPTER 8

1. Jeffrey Simpson, "The silence of big business on climate change is deafening," *Globe and Mail*, Jan. 27, 2007, A17.
2. John Wihbey, "United Nations IPCC fourth assessment report on climate change," *Journalist's Resource*, Aug. 3, 2011. https://journalistsresource.org/studies/environment/climate-change/united-nations-ipcc-report-climate-change.
3. "Stern Review on the Economics of Climate Change: Summary of Conclusions," Great Britain, HM Treasury, 2006. http://webarchive.nationalarchives.gov.uk/20130129110402/http://www.hm-treasury.gov.uk/sternreview_summary.htm.
4. Tyler Hamilton, "Climate change draws crowd," *Toronto Star*, Feb. 20, 2007, D1.
5. Don Butler, "The Stern message," *Ottawa Citizen*, Feb. 18, 2007, B6.
6. Andrew Hoffman and John Woody, *Climate Change: What's Your Business Strategy?* (Cambridge, MA: Harvard Business School Press, 2008), 1.
7. Vladimir Jankovic and Andrew Bowman, "After the green gold rush," *Economy and Society*, Vol. 43, Iss. 2, 2008, 235.
8. Geoffrey Scotton, "Rival climate plans costing Canada," *Calgary Herald*, June 25, 2008, D3.
9. Canadian Council of Chief Executives, Task Force on Environmental Leadership, "Clean Growth: Building a Canadian Environmental Superpower," Oct. 1, 2007. http://www.ceocouncil.ca/wp-content/uploads/archives/Clean_Growth_ELI_Policy_Declaration_October_1_2007.pdf.
10. *Ibid.*
11. Geoffrey Scotton, *op. cit.*
12. Tom d'Aquino, "New approach needed to deal with carbon," *National Post*, June 26, 2008, FP15.
13. Ross Beaty, Richard Lipsey and Stewart Elgie, "The shocking truth about B.C.'s carbon tax," *Globe and Mail*, July 9, 2014, A9.
14. Derrick Penner, "Businesses applaud efforts to reduce taxes in budget," *Vancouver Sun*, Feb. 20, 2008, A5.
15. "We have a winner," *Economist*, July 23, 2011, 35.
16. "World Bank Group president Jim Yong Kim speech at the Council on Foreign Relations," Washington, D.C., Dec. 8, 2014. http://www.worldbank.org/en/news/speech/2014/12/08/transcript-world-bank-group-president-jim-yong-kims-speech-qa-climate-change-solution.
17. "The climate challenge: Achieving zero emissions." Lecture by the OECD Secretary-General, Mr. Angel Gurria, London, Oct. 9, 2013. http://www.oecd.org/env/the-climate-challenge-achieving-zero-emissions.htm.
18. Toby A.A. Heaps, "Why Campbell's carbon tax gamble proved a good bet," *Corporate Knights*, June 3, 2015. http://www.corporateknights.com/channels/leadership/campbell-gamble-14333004/.
19. Justine Hunter, "How Beijing set off a Premier's smoke alarm," *Globe and Mail*, Feb. 23, 2008, A3.
20. David Beers, "Think Gordon Campbell is a global warming guru? Read on," *Globe and Mail*, Jan. 22, 2009, A17.
21. "Environmental groups disagree with B.C. NDP's stand on carbon tax," CBC News, Apr. 13, 2009. http://www.cbc.ca/news/canada/british-columbia/environmental-groups-disagree-with-b-c-ndp-s-stand-on-carbon-tax-1.790242.
22. Toby Heaps, *op. cit.*
23. Mark Hume, "Will B.C. Premier's win bring the carbon tax back to life?" *Globe and Mail*, May 14, 2009, A1.

24. Colleen Kimmett, "Campbell's carbon tax praised in Copenhagen," *The Tyee*, Dec. 16, 2009. https://thetyee.ca/Blogs/TheHook/Environment/2009/12/16/TaxPraised/.
25. Geoff Dembicki, "Gordon Campbell slammed by former backer, eco-activist Tzeporah Berman," *The Tyee*, Oct. 18, 2011. https://thetyee.ca/News/2011/10/18/Campbell-Slammed-By-Berman/.
26. Climate Action Network, "Dirty oil diplomacy," 2012. http://climateactionnetwork.ca/wp-content/uploads/2012/03/CAN_Dirty_Oil_Diplomacy.pdf.,
27. Murray Lyons, "Sask. must be productive on power play," *Saskatoon Star-Phoenix*, Mar. 7, 2008, C8; Frances Bula, "The man mapping out B.C.'s new energy plan," *Vancouver Sun*, Feb. 17, 2007, L4; "Throne speech a call to action," *Chilliwack Progress*, Feb. 20, 2007, A8.
28. Mike De Souza, "Harper's anti-Kyoto letter fuels Liberal counterattack," *Ottawa Citizen*, Jan. 31, 2007, A3.
29. Peter Newman, *Titans: How the New Canadian Establishment Seized Power* (Toronto: Viking, 1998), 446.
30. Andrew Mcintosh, "Stephen Harper," *Canadian Encyclopedia*, Jan. 27, 2016. http://www.thecanadianencyclopedia.ca/en/article/stephen-joseph-harper/.
31. Deirdre McMurdy, "Meet the PMO's chief trouble-shooter," *Ottawa Citizen*, Oct. 17, 2006, A4.
32. Greg Weston, "Who the heck shot Rona Ambrose?" *Whitehorse Star*, Nov. 6, 2006, 8.
33. "Liberals slam Harper aide's plum appointment," *Canadian Press*, Aug. 15, 2008.
34. Andrew Nikiforuk, "Bruce Carson scandal greased by Harper's oil sands agenda," *The Tyee*, Apr. 27, 2011. https://thetyee.ca/News/2011/04/27/CarsonOilSands/.
35. Charles Frank, "Oilsands face uphill battle in PR war," *Calgary Herald*, Aug. 16, 2008, D1.
36. Geoffrey Scotton, "Research institute gets new director," *Calgary Herald*, Aug. 15, 2008, B4.
37. Paul Wells, *The Longer I'm Prime Minister* (Random House Canada, 2013), 114.
38. "Canada pulls out of Kyoto Protocol," CBC News, Dec. 12, 2011. http://www.cbc.ca/news/politics/canada-pulls-out-of-kyoto-protocol-1.999072.
39. "Environment panel's end blamed on support for carbon tax," CBC News, May 15, 2012. http://www.cbc.ca/news/politics/environment-panel-s-end-blamed-on-support-for-carbon-tax-1.1164935.
40. Tim Harper, "Tories killing off voices, one at a time," *Toronto Star*, May 16, 2012, A6.

CHAPTER 9

1. "The Council of Canadians calls for a national energy strategy," Council of Canadians, June 30, 2011. https://canadians.org/fr/node/7483.
2. Romina Marino, "Canada needs national energy strategy to remain industry leader," *Canadian Press*, Mar. 23, 2006.
3. Marci McDonald, "The man behind Stephen Harper," *The Walrus*, Oct. 12, 2004. https://thewalrus.ca/the-man-behind-stephen-harper/; Donald Gutstein, *Not a Conspiracy Theory* (Toronto: Key Porter, 2009), 150–59.
4. Roger Gibbins, "Getting it right: A Canadian energy strategy for a carbon-constrained future," Calgary, Canada West Foundation, Nov. 2007. http://cwf.ca/wp-content/uploads/2015/11/CWF_GettingRight_EnergyStrategy_Report_NOV2007.pdf.
5. Max Bell Foundation, "Grants database, Getting it right," no date. http://www.maxbell.org/policy-intelligence/getting-it-right-balancing-climate-change-and-energy-policies-western-canada.
6. Roger Gibbins and Kari Roberts, "West should lead debate," *Calgary Herald*, Nov. 13, 2007, A12.
7. Karen Kleiss, "Canadian energy strategy key to Alberta's future," *Calgary Herald*, Jan. 16, 2012, A1.
8. Peter Newman, *Titans: How the New Canadian Establishment Seized Power* (Toronto: Viking, 1998), 461.
9. Linda McQuaig, *Holding the Bully's Coat* (Doubleday Canada, 2007), 225–26;

"Attendance list North American Forum," The New World Disorder, Sep. 21, 2006. http://www.wnd.com/2006/09/38017.

10. "About the Winnipeg Consensus," International Institute for Sustainable Development, 2011. http://www.winnipegconsensus.org/about.aspx.

11. Other key EPIC personnel at Banff were:
    - David Emerson, had been a deputy minister of Finance in B.C., and minister of International Trade in the Harper government. As a former vice-chair of the CEO Council, Emerson was close to d'Aquino. He gave the keynote address in Banff.
    - Gerry Protti had been an assistant deputy minister of Alberta Energy and then joined the industry to work mainly at EnCana Corporation. He was the founding president of CAPP, and became chair of the Alberta Energy Regulator, blurring the line between government and industry.
    - Doug Black was a leading oil and gas lawyer elected as an Alberta senator and appointed to the Senate by Stephen Harper in 2013. Black was a senior Alberta Conservative Party fundraiser. He was chief lobbyist for the Canadian Coalition for Responsible Environmental Solutions, whose purpose was to derail Kyoto ratification.
    - Larry Clausen provided EPIC's communication strategy. He worked in public relations for the energy industry for 25 years, and during the EPIC years was executive vice-president for PR firm Cohn & Wolfe in Calgary.
    - Dan Gagnier had been chief of staff at various times to David Peterson in Ontario and Jean Charest in Quebec and provided Liberal connections.

12. Jake Wright and Daniel LeBlanc, "Former Harper adviser Bruce Carson facing new charges," Globe and Mail, May 12, 2014. https://www.theglobeandmail.com/news/politics/former-harper-adviser-bruce-carson-facing-new-charges/article18611119/.

13. Laura Payton, "Bruce Carson, former PMO staffer, has banking records seized by RCMP," CBC News, Mar. 5, 2014. http://www.cbc.ca/news/politics/bruce-carson-former-pmo-staffer-has-banking-records-seized-by-rcmp-1.2561271.

14. Tony Spears, "PM's ex-senior advisor 'secret sauce' for energy lobby group," Ottawa Sun, Mar. 6, 2014. https://ottawasun.com/2014/03/06/pms-ex-senior-advisor-secret-sauce-for-energy-lobby-group/.

15. Kathleen Harris, "Bruce Carson likely will avoid jail time for influence peddling," CBC News, May 3, 2018. https://www.cbc.ca/news/politics/carson-sentence-influence-peddling-harper-1.4646710.

16. Cited in Kevin Taft, Oil's Deep State (Toronto: Lorimer, 2017), 30.

17. Natural Resources Canada, "Canadian energy ministers commit to further innovation and strengthen collaboration," Sept. 17, 2010. https://www.nrcan.gc.ca/media-room/news-release/74/2010-09/2797.

18. Shawn McCarthy, "National energy strategy gains clout," Globe and Mail, July 11, 2011, B1.

19. Canadian Council of Chief Executives, "Kananaskis 2011: Building an agenda for a sound energy future," July 2011. http://thebusinesscouncil.ca/publications/kananaskis-2011-building-an-agenda-for-a-sound-energy-future-submission-to-the-conference-of-energy-and-mines-ministers/.

20. Natural Resources Canada, "Energy ministers support collaborative approach to energy," July 19, 2011. https://www.nrcan.gc.ca/media-room/news-release/2011/2421.

21. Jason Fekete, "Candidates square off on energy, environment," Calgary Herald, June 23, 2011, A6.

22. Daniel Proussalidis, "Thumbs down to NDP and Quebec: Alberta premier," canoe.com, Nov. 17, 2011. http://cnews.canoe.com/CNEWS/Politics/2011/11/17/18983191.html.

23. Canada's Premiers, "Premiers guide development of Canada's energy resources," Halifax, July 27, 2012. http://www.canadaspremiers.ca/en/latest-news/15-2012/138-premiers-guide-development-of-canada-s-energy-resources.

24. Energy Policy Institute of Canada, "A Canadian energy strategy framework," Aug. 2012. https://issuu.com/energypolicyinstitutecanada/docs/canadianenergystrategy/67.

25. ForestEthics Advocacy, "Who writes the rules?" Oct. 3, 2016, 2. https://issuu.com/stand.earth/docs/who_writes_the_rules.

26. *Ibid.*, 3.
27. Business Council of Canada, "John R. Dillon," no date. http://thebusinesscouncil.ca/about-the-council/council-staff/.
28. "National survey on energy literacy: Public trust and confidence," *Alberta Oil*, Feb. 2, 2015. https://www.albertaoilmagazine.com/2015/02/public-trust-confidence/.
29. Kevin Taft, "Best interests of Albertans, or big oil?" *Edmonton Journal*, Apr. 29, 2015, A19.
30. Jennifer Brown, "BD&P founder was business leader and philanthropist," *Canadian Lawyer*, Sept. 3, 2013. https://www.canadianlawyermag.com/legalfeeds/author/jennifer-brown/bdandp-founder-was-business-leader-and-philanthropist-5314/.
31. Annalise Klingbeil and Matt McClure, "University of Calgary president, who intervened in research centre funded by Enbridge, defends her role despite sitting on related board," *Financial Post*, Nov. 3, 2015. http://business.financialpost.com/news/energy/enbridge-university-of-calgary.
32. Dale Eisler, "Energy literacy in Canada: A summary," *School of Public Policy, Research Papers*, Vol. 9, Issue 1 (Jan. 2016). https://www.policyschool.ca/wp-content/uploads/2016/03/energy-literacy-canada-eisler.pdf.
33. *Ibid.*
34. David Hackett, *et al.*, "Pacific Basin Heavy Oil Refining Capacity," *School of Public Policy, Research Papers*, Vol. 6, Issue 8 (Feb. 2013). https://www.policyschool.ca/wp-content/uploads/2016/03/pacific-basin-hackett-noda-grissom-moore-winter.pdf; Bill Graveland, "Report Says Time Running Out for Canadian Oil Producers to Access Pacific Rim," *Canadian Press*, Feb. 6, 2013.
35. Jennifer Winter and Michal Moore, "The 'Green Jobs' Fantasy: Why the Economic and Environmental Reality Can Never Live Up to the Political Promise," *School of Public Policy, Research Papers*, Vol. 6, Issue 31 (Oct. 2013). https://www.policyschool.ca/wp-content/uploads/2016/03/j-winter-green-jobs-final.pdf; Jennifer Winter, "The Myth of Green-Job Creation," *Vancouver Sun*, Oct. 28, 2013, A10.
36. Alan Gelb, "Should Canada Worry About a Resource Curse?" *School of Public Policy, Research Papers*, Vol. 7, Issue 2 (Jan. 2014). https://www.policyschool.ca/wp-content/uploads/2016/03/resource-curse-gelb.pdf.
37. Kenneth McKenzie, "Make the Alberta carbon levy revenue neutral," School of Public Policy, Briefing Paper, Vol. 9, Issue 15 (April 2016). http://www.policyschool.ca/wp-content/uploads/2016/05/carbon-levy-revenue-neutral-mckenzie.pdf.
38. Clive Mather, Nancy Olewiler, and Stewart Elgie, "B.C.'s Carbon Tax Shift Is Smart Public Policy," *Globe and Mail*, Nov. 29, 2007, A21.
39. Sustainable Prosperity, "What We Do: Introducing Sustainable Prosperity : A Think Tank. A Do Tank," 2013. https://web.archive.org/web/20131006090601/http://www.sustainableprosperity.ca/What+We+Do+EN..
40. Elizabeth Thompson, "Former Liberal campaign co-chair was being paid by TransCanada," Oct. 15, 2015. http://ipolitics.ca/2015/10/15/trudeau-distances-himself-from-former-campaign-co-chair/.
41. Information about McColl's lobbying activities was retrieved from the Office of the Commissioner of Lobbying of Canada, Mar. 10, 2016. https://lobbycanada.gc.ca.
42. Glen McGregor, "Liberal lobbyists boast close connections to Trudeau government," *Ottawa Citizen*, Dec. 16, 2015. http://ottawacitizen.com/news/politics/the-garggyle-liberal-lobbyists-boast-close-connections-to-trudeau-government.
43. Brian Murray and Nicholas Rivers, "British Columbia's revenue-neutral carbon tax," *Energy Policy*, Vol. 86, 2015, 674–683.
44. *Ibid.*, 682.
45. Marvin Shaffer, "What's wrong with a revenue neutral carbon tax?" *Policy Note*, Jan. 22, 2016. http://www.policynote.ca/whats-wrong-with-a-revenue-neutral-carbon-tax/.
46. Werner Antweiler, "Is BC's carbon tax salient?" *Werner's Blog*, June 15, 2015. https://wernerantweiler.ca/blog.php?item=2015-06-15.

47. "The impacts of the carbon tax on vehicle fuels in Metro Vancouver," *Pacific Analytics*, Mar. 2015. http://pacificanalytics.ca/node/46.
48. Elgie and McClay, 2013: Stewart Elgie Sustainable Prosperity founder and chair, Jessica McClay Sustainable Prosperity research assistant; Rivers and Schaufele, 2012: Nic Rivers co-chair Sustainable Prosperity low carbon economy program, Brandon Schaufele former Sustainable Prosperity research director; Gulati and Gholami, 2015: Zahra Gholami: did thesis with Sustainable Prosperity funding; Beck *et al.*, 2015: Marisa Beck co-wrote with co-author Randall Wigle a Sustainable Prosperity paper on carbon pricing; Bernard *et al.*, 2014: Jean-Thomas Bernard principal investigator of SP project linking natural capital and productivity.
49. Marc Lee, "Don't believe the hype on BC's carbon tax," CCPA Policy Note, Mar. 2, 2016. http://www.policynote.ca/dont-believe-the-hype-on-bcs-carbon-tax.
50. Marc Lee, "The case against a revenue-neutral carbon tax," *The Progressive Economics Forum*, Jan. 15, 2015. http://www.progressive-economics.ca/2015/01/15/the-case-against-a-revenue-neutral-carbon-tax/.

CHAPTER 10

1. "Environmental pioneer Maurice Strong mourned at COP21," *Environment News Service*, Dec. 8, 2015. http://ens-newswire.com/2015/12/08/environmental-pioneer-maurice-strong-mourned-at-cop21/.
2. "Maurice Strong remembered as father of world environmental movement," *Canadian Press*, Jan. 7, 2016.
3. Laura Serra, "Maurice Strong's birthday," *Globe and Mail*, May 3, 2014, M2.
4. Peter Foster, "Breaking fake eco news," *National Post*, July 26, 2017, FP9.
5. Peter Christoff, "The promissory note: COP21 and the Paris Climate Agreement," *Environmental Politics*, Vol. 25, No. 5, 2016, 767.
6. Nick Buxton, "COP21 charades: Spin, lies and real hope in Paris," *Globalizations*, Vol. 13, No. 6, 2016, 934.
7. Nat Keohane, "Report back from Paris: What the new climate deal means — and where we go from here," Environmental Defense Fund, Dec. 15, 2015. https://www.edf.org/blog/2015/12/15/report-back-paris-what-new-climate-deal-means-and-where-we-go-here.
8. David Suzuki with Steve Kux, "Paris Agreement marks a global shift for climate," *David Suzuki Foundation*, Dec. 17, 2015. http://www.davidsuzuki.org/blogs/science-matters/2015/12/paris-agreement-marks-a-global-shift-for-climate/.
9. Luke Kemp, "A systems critique of the 2015 Paris Agreement on climate," in Moazzem Hossain, Robert Hales and Tapan Sarker, eds., *Pathways to a Sustainable Economy* (Cham, Switzerland: Springer, 2018), 31.
10. Daniel Schwartz, "'Worthless words' or 'diplomatic triumph?' Climate scientists weigh in on Paris agreement," CBC News, Dec. 21, 2015. http://www.cbc.ca/news/technology/climate-scientists-paris-1.3366751.
11. Tom Bawden, "COP21: Paris deal far too weak to prevent devastating climate change, academics warn," *The Independent*, Jan 9, 2016. http://www.independent.co.uk/environment/climate-change/cop21-paris-deal-far-too-weak-to-prevent-devastating-climate-change-academics-warn-a6803096.html.
12. Oliver Milman, "James Hansen, father of climate change awareness, calls Paris talks 'a fraud,'" *The Guardian*, Dec. 12, 2015. https://www.theguardian.com/environment/2015/dec/12/james-hansen-climate-change-paris-talks-fraud.
13. Bill McKibben, "Falling short on climate," *New York Times*, Dec. 14, 2015, 23.
14. Daniel Schwartz, *op. cit.*
15. Clive Spash, "This changes nothing: The Paris Agreement to ignore reality," *Globalizations*, Vol. 13, No. 6, 2016, 930.
16. Lara Marlowe, "Energy industry accused of faking green credentials as divisions emerge in Paris," *Irish Times*, Dec. 4, 2015, 10.

17. "Climate negotiations and fossil fuel industry," *The Financial Express* (Bangladesh), Sept. 21, 2016.

18. Michael Slezak, "Marrakech climate talks: giving the fossil fuel lobby a seat at the table," *The Guardian*, Nov. 6, 2016. https://www.theguardian.com/environment/2016/nov/07/marrakech-climate-talks-giving-the-fossil-fuel-lobby-a-seat-at-the-table.

19. "Impact assessment of the WHO FCTC: Report by the expert group," July 27, 2016. http://www.who.int/fctc/cop/cop7/FCTC_COP_7_6_EN.pdf.

20. "Framework Convention on Tobacco Control," World Health Organization, 2005. http://apps.who.int/iris/bitstream/10665/42811/1/9241591013.pdf.

21. "Guidelines for implementation of Article 5.3 of the WHO Framework Convention on Tobacco Control," World Health Organization, no date. http://www.who.int/fctc/guidelines/article_5_3.pdf.

22. "Business Dialogue meeting with Ségolène Royal and Gérard Mestrallet," Engie, Feb. 16, 2016. https://www.youtube.com/watch?v=iMnT_HN5ARU.

23. Ed King, "Stop demonising oil and gas companies, says UN climate chief," Climate Change News, May 26, 2015. http://www.climatechangenews.com/2015/05/26/stop-demonising-oil-and-gas-companies-says-un-climate-chief/.

24. John Moorhead and Tim Nixon, "Global 500 greenhouse gases performance 2010–2015," *Thomson Reuters*, June 2016, 3. https://www.thomsonreuters.com/content/dam/openweb/documents/pdf/corporate/Reports/global-500-greenhouse-gases-performance-2010-2015.pdf.

25. Sarah Benabou, Nils Moussu and Birgit Muller, "The business voice at COP21," in Stefan Aykut, Jean Foyer and Edouard Morena, eds., *Globalising the Climate: COP21 and the Climatisation of Global Debates* (New York: Routledge, 2017), 63.

26. Damian Carrington, "US and Chinese companies dominate list of most-polluting coal plants," *The Guardian*, Mar. 13, 2015. https://www.theguardian.com/environment/2015/mar/13/us-and-chinese-companies-dominate-list-of-most-polluting-coal-plants.

27. Ben Caldecott, Gerard Dericks and James Mitchell, "Stranded assets and sub-critical coal," Smith School of Enterprise and the Environment, University of Oxford, Mar. 2015. http://www.smithschool.ox.ac.uk/research-programmes/stranded-assets/Stranded%20Assets%20and%20Subcritcial%20Coal%20-%20The%20Risk%20to%20Investors%20and%20Companies%20-%20April15.pdf.

28. "GDF Suez becomes Engie," Engie, Apr. 24, 2015. http://www.engie.com/en/journalists/press-releases/gdf-suez-becomes-engie/.

29. "Engie's new advertising campaign: From the name change to a gateway of solutions," Engie, Sept. 25, 2015. http://www.engie.com/en/journalists/press-releases/engies-new-advertising-campaign-a-gateway-of-solutions/.

30. "French energy giant to stop new coal," Friends of the Earth Europe, Oct. 15, 2015. http://foeeurope.org/French-energy-giant-stop-new-coal-151015.

31. "CO2 capture and storage," Engie, no date. http://www.engie.com/en/commitments/reasearch-innovation/carbon dioxide-capture-storage/.

32. "Riyadh's for solar and other renewables," *APS Review Downstream Trends*, Dec. 26, 2016, Vol. 87, No. 26.

33. "Oil and gas majors call for carbon pricing," Shell Global, June 1, 2015. http://www.shell.com/media/news-and-media-releases/2015/oil-and-gas-majors-call-for-carbon-pricing.html.

34. Article 6.2, Paris Agreement, United Nations, Dec. 12, 2015. https://unfccc.int/files/essential_background/convention/application/pdf/english_paris_agreement.pdf.

35. "COP21: Carbon pricing text agreed in climate deal," *The Environmentalist*, Dec. 9, 2015. https://www.environmentalistonline.com/article/cop21-carbon-pricing-text-agreed-climate-deal.

36. Sarah Benabou *et al.*, *op. cit.*, 60.

37. *Ibid.*, 61.

38. "International Emissions Trading Association Paris COP21 summary," Dec, 14, 2015, 3. http://www.ieta.org/resources/Conferences_Events/COP21/International Emissions Trading Association_COP_21_Summary.pdf.
39. Sarah Benabou *et al.*, *op. cit.*, 69.
40. Article 4.4, Paris Agreement, *op. cit.*
41. "Climate deal typo hiccup solved 'in a small room,' France says," *Reuters*, Dec. 15, 2015. http://www.reuters.com/article/us-climatechange-summit-error-idUSKBN0TY1CS20151215.
42. Stefan Aykut and Monica Castro, "The end of fossil fuels?" in Stefan Aykut, Jean Foyer and Edouard Morena, eds., *Globalising the Climate: COP21 and the Climatisation of Global Debates* (New York: Routledge, 2017), 182–83.
43. "Paul Polman praises historic Paris Agreement," Unilever, Dec. 12, 2015. https://www.unilever.com/news/news-and-features/Feature-article/2015/Polman-praises-Paris-Agreement-121215.html.
44. Article 4.2, Paris Agreement, *op. cit.*
45. Robert Brulle, "Institutionalizing delay," *Climatic Change*, Vol. 122, No. 4, 2014, 690.
46. John Cook *et al.*, "Quantifying the consensus on anthropogenic global warming in the scientific literature," *Environmental Research Letters*, Vol. 8, No. 2, 2013.
47. Joe Morris, "Sad news about a tribune of liberty," *Illinois Review*, Oct. 4, 2011. http://illinoisreview.typepad.com/illinoisreview/2011/10/sad-news-about-a-tribune-of-liberty.html; Steve Horn, "Heartland Institute: a manifestation of the Kochtopus empire," DeSmog, Mar. 1, 2012. https://www.desmogblog.com/heartland-institute-manifestation-kochtopus-empire.
48. "Toxic shock," *The Economist*, May 26, 2012. http://www.economist.com/node/21555894.
49. Justin Gillis, "Clouds' effect on climate change is last bastion for dissenters," *New York Times*, Apr. 30, 2012. http://www.nytimes.com/2012/05/01/science/earth/clouds-effect-on-climate-change-is-last-bastion-for-dissenters.html.
50. Leo Hickman, "Heartland Institute compares belief in global warming to mass murder," *The Guardian*, May 4, 2012. https://www.theguardian.com/environment/blog/2012/may/04/heartland-institute-global-warming-murder.
51. Kyla Mandel, "Heartland Institute kicks journalists out of 'public' climate denial event in Paris," DeSmog, Dec. 7, 2015. https://www.desmogblog.com/2015/12/07/heartland-institute-kicks-journalists-out-public-climate-denial-press-conference-paris.
52. See, for instance, Fred Smith, Jr., "Eco-Socialism: Threat to liberty around the world," Competitive Enterprise Institute, Sept. 20, 2003. http://cei.org/pdf/3818.pdf.
53. Fred Smith, Jr., "Markets and the environment: a critical reappraisal," *Contemporary Economic Policy*, Vol. 13, No. 1, Jan. 1995, 62–73.
54. "Myron Ebell," DeSmog, no date. https://www.desmogblog.com/myron-ebell.
55. Jennifer Dlouhy, ""EPA nominee vows to limit body's reach," *Vancouver Sun*, Jan. 19, 2017, C3.
56. Brian Kennedy, "Clinton, Trump supporters worlds apart on views of climate change and its scientists," Pew Research Center, Oct. 10, 2016. http://www.pewresearch.org/fact-tank/2016/10/10/clinton-trump-supporters-worlds-apart-on-views-of-climate-change-and-its-scientists/.
57. Constantine Boussalis and Travis Coan, "Text-mining the signals of climate change doubt," *Global Environmental Change*, Vol. 36, 2016, 98.
58. "Letter to the Honorable Donald J. Trump," May 8, 2017. http://americanenergyalliance.org/wp-content/uploads/2017/05/Letter-to-President-Trump-on-Paris-Climate-Treaty-from-Competitive Enterprise Institute-American Energy Alliance-and-39-organizations-8-May-2017.pdf.
59. John Siciliano, "Dozens of groups press Trump to exit Paris climate deal," *Washington Examiner*, May 8, 2017. http://www.washingtonexaminer.com/dozens-of-groups-press-trump-to-exit-paris-climate-deal/article/2622395.

60. "Businesses urge president to remain in Paris Agreement," Center for Climate and Energy Solution, no date. https://www.c2es.org/international/business-support-paris-agreement.

61. Larry Light, "Why US businesses said 'stay in the Paris accord,'" CBS News MoneyWatch, June 2, 2017. https://www.cbsnews.com/news/paris-climate-agreement-us-corporate-support/.

62. Matt Egan, "Exxon to Trump: Don't ditch Paris climate change deal," Mar. 29, 2017. http://money.cnn.com/2017/03/29/investing/exxon-trump-paris-climate-change/index.html.

63. Larry Light, *op. cit.*

64. Philip Mirowski, *Never Let a Serious Crisis Go To Waste* (London: Verso, 2013), 337.

65. *Ibid.*, 338.

66. Raymond Pierrehumbert, "Climate hacking is barking mad," *Slate*, Feb. 10, 2015. http://www.slate.com/articles/health_and_science/science/2015/02/nrc_geoengineering_report_climate_hacking_is_dangerous_and_barking_mad.html.

67. Quoted in Katherine Ellison, "Why climate change skeptics are backing geoengineering," *Wired*, Mar. 28, 2018. https://www.wired.com/story/why-climate-change-skeptics-are-backing-geoengineering/.

68. Nick Stockton, "Climate change is here. It's time to talk about geoengineering," *Wired*, July 20, 2017. https://www.wired.com/story/lets-talk-geoengineering/.

69. Katherine Ellison, *op. cit.*; Peter Coy, "It's already too late," *BusinessWeek*, Nov. 30, 2015, 58.

70. Shawn McCarthy, "Bill Gates-funded B.C. startup says it can slash carbon-capture costs, replace gasoline at competitive price," *Globe and Mail*, June 7, 2018. https://www.theglobeandmail.com/business/article-bc-start-up-says-it-can-cut-carbon-capture-costs-dramatically/.

CHAPTER 11

1. Chris Varcoe, "When the oilsands hit paydirt," *Calgary Herald*, Sept. 26, 2017, A6.

2. C.V. Deutsch and J.A. McLennan, "Guide to SAGD (steam assisted gravity drainage) reservoir characterization using geostatistics," Centre for Computational Geostatistics, University of Alberta, 2005. https://web.archive.org/web/20081209030733/http://www.uofaweb.ualberta.ca/ccg/pdfs/Vol3-IntroSAGD.pdf.

3. Statistics are from "Statistical handbook for Canada's upstream petroleum industry," Canadian Association of Petroleum Producers, Feb. 2018, Table 3.2. https://www.capp.ca/publications-and-statistics/.

4. Chris Varcoe, *op. cit.*

5. "Oil Sands: Cold Lake," Alberta Culture and Tourism, no date. http://www.history.alberta.ca/energyheritage/sands/underground-developments/in-situ-development/cold-lake.aspx.

6. Chris Varcoe, "Foreign exodus follows investment stampede," *Calgary Herald*, Sept. 29, 2017, A6.

7. Scott Anderson and Andrew Culbert, "How TransCanada and 'dark money' groups pumped millions into Keystone XL," CBC News Nov. 8, 2017. http://www.cbc.ca/news/business/transcanada-dark-money-keystone-xl-1.4384440.

8. Derek Burney and Eddie Goldenberg, "Shipping Alberta's oil to Asia? The route may lie east, not west," *Globe and Mail*, Dec. 13, 2011, A17.

9. Carrie Tait, "Total offers a cautionary tale on the perils of oilsands projects," *Globe and Mail*, May 31, 2014, B3.

10. Claudia Cattaneo, "Big Oil's kindred spirit," *National Post*, May 17, 2014, FP3.

11. Shawn McCarthy, "The NDP's oil sands balancing act," *Globe and Mail*, May 9, 2015, B1.

12. Carrie Tait and Jeff Lewis, "Canadian Natural's stance with NDP could backfire, analysts warn," *Globe and Mail*, May 28, 2015. https://beta.theglobeandmail.com/report-on-business/industry-news/energy-and-resources/cnrl-cancels-investor-day-seeks-clarity-on-ndp-energy-policies/article24678849/.

13. Kyle Bakx, "CNR blames NDP for financial loss," CBC News, Aug. 6, 2015. http://www.cbc.ca/news/business/canadian-natural-resources-blames-ndp-for-financial-loss-1.3181541.

14. Kelly Sinoski," Major Imperial shareholder held private fundraiser for Clark in Calgary," *Vancouver Sun*, Aug. 9, 2014, A3.
15. Cited in Kevin Taft, *Oil's Deep State* (Toronto: Lorimer, 2017), 171.
16. Andrew Leach, "Are oil sands incompatible with action on climate change?" *Maclean's*, Jan. 10, 2015. http://www.macleans.ca/economy/economicanalysis/are-oil-sands-incompatible-with-action-on-climate-change/.
17. Christophe McGlade and Paul Ekins, "The geographical distribution of fossil fuels unused when limiting global warming to 2° C," *Nature*, Vol. 517, Jan. 8, 2015, 190.
18. Andrew Leach, "We cannot just sit and wait for the next oil boom," *Globe and Mail*, Jan, 23, 2015, A10.
19. Jason Markusoff, "Shannon Phillips, Alberta's Minister of hard hits," *Maclean's* Jan. 11, 2016. http://www.macleans.ca/news/canada/shannon-phillips-albertas-minister-of-hard-hits/.
20. Lauren Krugel, "General public to have say in Alberta climate change policy review," *Canadian Press*, Aug. 14, 2015.
21. Jeffrey Jones, "An unlikely carbon accord: How two sides became one," *Globe and Mail*, Nov. 25, 2015, B1.
22. Max Fawcett, "The artist of the deal," *Report on Business Magazine*, Nov. 22, 2016. https://beta.theglobeandmail.com/report-on-business/rob-magazine/the-elusive-billionaire-who-wants-to-make-the-oil-sands-greener/article32955602/.
23. Don Braid, "Notley gives Alberta climate-change jolt," *Calgary Herald*, Nov. 23, 2015, A3.
24. Stephen Ewart, "New climate plan touted as a PR tool," *Calgary Herald*, Nov. 23, 2015, A4.
25. "Alberta's climate leadership plan," Your Alberts, Nov. 22, 2015. https://www.youtube.com/watch?v=WkwlLWfVag8&t=584s.
26. Max Fawcett, *op. cit.*
27. Climate Change Advisory Panel, "Climate leadership report to minister," Government of Alberta, 2015, 41. https://www.alberta.ca/documents/climate/climate-leadership-report-to-minister.pdf.
28. *Ibid.*, 37.
29. *Ibid.*, 33.
30. *Ibid.*, 40.
31. *Ibid.*, 33.
32. Chris Varcoe, "Carbon tax is on the table: Province expects to release elements of climate change plan before talks," *Calgary Herald*, Nov. 14, 2015, A3.
33. Climate Change Advisory Panel, *op. cit.*, 29.
34. *Ibid.*
35. Dale Beugin *et al.*, "Comparing stringency of carbon pricing policies," Canada's Ecofiscal Commission, Jan. 2017, 6. http://ecofiscal.ca/wp-content/uploads/2016/07/Ecofiscal-Commission-Comparing-Stringency-Carbon-Pricing-Report-July-2016.pdf.
36. Don Drummond, Nancy Olewiler and Christopher Ragan, "Carbon price vs. regulations: The better choice is clear," *Globe and Mail*, Oct. 5, 2016, B4.
37. Reid Southwick, "TransAlta moving ahead with hydro project after deal on coal shutdown," *Calgary Herald*, Nov. 26, 2016, A4; Armina Ligaya, "Capital Power earns upgrade," *National Post*, Nov. 26, 2016, FP6.
38. Shareholders disagreed and voted 53 per cent against a non-binding resolution on the executive compensation package at the company's annual general meeting. "TransAlta commits to phase out coal power years ahead of government deadline," see *Canadian Press*, Apr. 20, 2017.
39. John Platt, "What is methane and why should you care?" *Mother Nature Network*, May 11, 2017. https://www.mnn.com/earth-matters/energy/stories/what-is-methane-and-why-should-you-care.
40. "Tougher methane reduction measures in Alberta sought by environmental groups," CBC News, June 28, 2018. https://www.cbc.ca/news/canada/calgary/alberta-methane-reduction-environmental-groups-1.4727160.

41. Stephen Ewart, *op.cit.*
42. Quoted in Jason Markusoff, "Ka-boom or ka-bust? Reaction to Alberta's climate change plan," *Maclean's*, Nov. 23, 2015.
43. Eric Reguly, "Alberta carbon plan is deceptive," *Globe and Mail*, Nov. 28, 2015, B1.
44. Claudia Cattaneo, "Producer backlash grows over NDP oilsands cap," *National Post*, Dec. 10, 2015, FP1.
45. Geoffrey Morgan, "Carbon taxes, fees divide oilpatch association," *Calgary Herald*, June 21, 2017, B1.

CHAPTER 12

1. Mark Burgess, "Alberta election upends lobbying industry, will lead to 'buying season:' consultants," *Hill Times*, May 18, 2015. https://www.hilltimes.com/2015/05/15/alberta-election-upends-lobbying-industry-will-lead-to-buying-season-consultants/32181.
2. Steven Chase, "Leading party strategists join forces," *Globe and Mail*, Feb. 8, 2013, A8.
3. When Marcella Munro joined the firm in August 2017, Ken Boessenkool wrote on KTG's LinkedIn page, "Brian Topp, Don Guy and I welcome Marcella Munro as Senior Strategist at KTG Public Affairs." Ken Boessenkool, "Welcome Marcella Munro; Bon voyage Jamey Heath," KTG Public Affairs, Aug. 11, 2017. https://www.linkedin.com/pulse/welcome-marcella-munro-bon-voyage-jamey-health-ken-boessenkool/.
4. TransCanada has been one of the most persistent energy industry lobbyists in Alberta and Ottawa. See the profile: Mehreen Amani Khalfan, "Unplugging the dirty energy economy," Ottawa, Polaris Institute, June 2015. https://d3n8a8pro7vhmx.cloudfront.net/polarisinstitute/pages/109/attachments/original/1433444082/TransCanada_Profile.pdf.
5. "TransCanada Corporation," Alberta Lobbyist Registry, Aug. 11, 2018. https://www.albertalobbyistregistry.ca.
6. Simon Doyle, "Conservative lobbyists join Harper's war room ahead of election," *Globe and Mail*, Aug. 6, 2015. https://www.theglobeandmail.com/news/politics/globe-politics-insider/conservative-lobbyists-join-harpers-war-room-ahead-of-election/article25870011/.
7. Don Braid, "NDP's new Calgary hand declares love for oilsands and her BMW," *Calgary Herald*, Oct. 6, 2015. http://calgaryherald.com/opinion/columnists/braid.
8. James Wood and Darcy Henton, "Prentice sees hope for Keystone in new Congress," *Calgary Herald*, Nov. 19, 2014, A3.
9. Sheila Pratt and William Marsden, "Pipeline veto not a worry," *Edmonton Journal*, Jan. 7, 2015, A1.
10. Les Whittington, "Pipeline politics end with NDP win," *Toronto Star*, May 8, 2015, A6.
11. Mark Burgess, *op. cit.*
12. "Pacific NorthWest LNG: Advocacy a case study," Wazuku Advocacy Group, no date. https://www.wazuku.ca/s/Wazuku_Case-Study_PNWLNG.pdf.
13. Mariam Ibrahim, "Power shift," *Edmonton Journal*, Sept. 20, 2015, B1.
14. John Cotter, "Long-time B.C. New Democrat hired by Alberta lobbying firm," *Globe and Mail*, June 8, 2015, S4.
15. Chris Varcoe, "Emissions cap coming for oilsands, but it may not matter," *Calgary Herald*, June 21, 2016. http://calgaryherald.com/business/energy/varcoe-emissions-cap-coming-for-oilsands-but-it-may-not-matter.
16. Dan Healing, "Use of natural gas in oilsands for crude called wasteful," *Vancouver Sun*, Apr. 20, 2017, C3.
17. Geoffrey Morgan, "Imperial eyes new $2-billion oilsands plant," *National Post*, Mar. 12, 2016, FP4.
18. Innovation, Science and Economic Development Canada, "Governments of Canada and Alberta invest in cutting-edge clean technologies to encourage clean growth," Nov. 3, 2017. http://www.newswire.ca/news-releases/governments-of-canada-and-alberta-invest-in-cutting-edge-clean-technologies-to-encourage-clean-growth-654964223.html.

19. "New technologies expected to boost profits and limit emissions from the oilsands," *Canadian Press*, Jan. 17, 2017.

20. Chris Varcoe, 'Study says oilsands can cut costs and emissions," *Calgary Herald*, Mar. 13, 2017, A4.

21. Kelly Cryderman, "Suncor for the first time, sets specific greenhouse gas reduction goal," *Globe and Mail*, July 23, 2016, B3.

22. Ian Bickis, "Latest XPrize offers $20 million to find new uses for carbon emissions," *Canadian Press*, Sept. 29, 2015.

23. Laura Kane, "Canadian teams reimagine carbon dioxide emissions for $20M competition," *Canadian Press*, Jan. 8, 2017.

24. Simon Doyle, "Oil-and-gas lobbyists get battle ready before the NDP's royalty review," *Globe and Mail*, May 19, 2015, B2.

25. Tracy Johnson, "Oilpatch wins the day in royalty review," CBC News, Jan. 29, 2016. http://www.cbc.ca/news/business/oilpatch-wins-day-royalties-1.3424045.

26. Kyle Bakx, "Alberta royalty review recommends few changes for oil and gas industry," CBC News, Jan. 29, 2016. http://www.cbc.ca/news/canada/calgary/alberta-royalty-review-changes-1.3424556.

27. James Wood, "Who's on the Alberta government royalty review panel," *Calgary Herald*, Aug. 28, 2015. http://calgaryherald.com/business/energy/whos-on-the-alberta-government-royalty-review-panel.

28. Andrew Nikiforuk, "Royalty miscalculation cost Alberta billions, expert says," *The Tyee*, May 2, 2015. https://thetyee.ca/News/2015/05/02/Royalty-Miscalculation-Cost-Alberta-Billions/.

29. Ricardo Alcuna, "The good, the bad, and the ugly of Alberta's new royalty framework," Parkland Institute, Feb. 1, 2016. http://www.parklandinstitute.ca/the_good_the_bad_and_the_ugly.

30. Rick Bell, "Alberta labour leader Gil McGowan pushes back against Premier Rachel Notley's royalty U-turn," *Calgary Sun*, Jan. 30, 2016. http://www.calgarysun.com/2016/01/30/alberta-labour-leader-gil-mcgowan-pushes-back-against-premier-rachel-notleys-royalty-u-turn.

31. Dan Healing, "Athabasca Oil praised for getting a bargain on Statoil oilsands assets," *Canadian Press*, Dec. 16, 2016.

32. Joe Carroll, "Oilsands region wipes out reserves of big explorers," *Vancouver Sun*, Feb. 24, 2017, C2. Under US Securities and Exchange Commission rules, proven reserves can include only oil and gas fields that can be produced economically within the next five years. These reserves can be added back if conditions change.

33. Jeffrey Jones, "From afar Murray Edwards makes his biggest splash," *Globe and Mail*, Mar. 10, 2017, B1.

34. "Marathon Oil announces $2.5 billion Canadian oil sands divestiture and $1.1 billion Permian Basin acquisition," Marathon Oil, Mar. 9, 2017. http://www.marathonoil.com/News/Press_Releases/Press_Release/?id=1016672.

35. Dan Healing, "Cenovus Energy buying most of ConocoPhillips's Canadian assets for C$17.7B," *Canadian Press*, Mar. 29, 2017.

36. Jeff Lewis, "Cenovus piles on debt for ConocoPhillips deal," *Globe and Mail*, Mar. 31, 2017, B1.

37. Jeff Lewis, "Oil sands earnings gaining momentum," *Globe and Mail*, Nov. 3, 2017, B1.

38. Brent Jang, "Shell-BG Group merger a game changer for B.C.'s LNG industry," *Globe and Mail*, Apr. 8, 2015. http://www.theglobeandmail.com/report-on-business/industry-news/energy-and-resources/shell-bg-group-merger-a-game-changer-for-bcs-lng-industry/article23853965/.

39. Kelly Cryderman, "Shell officially shelves plans to build Prince Rupert LNG project," *Globe and Mail*, Mar. 10, 2017. https://beta.theglobeandmail.com/report-on-business/industry-news/energy-and-resources/shell-officially-shelves-plans-to-build-prince-rupert-lng-project/article34272575.

40. Jon Harding, "Western seen as next Shell target," *National Post*, Mar. 19, 2007, FP1.

CHAPTER 13

1. Peter O'Neil, "The inside story of Kinder Morgan's approval," *Vancouver Sun*, Jan. 7, 2017, B1.
2. John Manley, "Trudeau wants to get serious about climate change. This is how he should start," *iPolitics*, Nov. 3. 2015. http://ipolitics.ca/2015/11/03/trudeau-wants-to-get-serious-about-climate-change-this-is-how-he-should-start/.
3. Claudia Cattaneo, "Business awaits Trudeau's pipelines 'grand bargain,'" *Calgary Herald*, Nov. 29, 2016, B1.
4. *Ibid.*
5. "Business leaders applaud pipeline decision," Business Council of Canada, Nov. 30, 2016. http://thebusinesscouncil.ca/news/statement-honourable-john-manley-president-ceo-business-council-canada-yesterdays-announcement-pipeline-projects/.
6. "Clean growth: Building a Canadian environmental superpower," Canadian Council of Chief Executives, Oct. 1, 2007. http://www.ceocouncil.ca/wp-content/uploads/archives/Clean_Growth_ELI_Policy_Declaration_October_1_2007.pdf; "Pan-Canadian framework on clean growth and climate change," Environment and Climate Change Canada, Dec. 2016. https://www.canada.ca/content/dam/themes/environment/documents/weather1/20170125-en.pdf.
7. Derek Leahy, "Trudeau's national climate meeting seen as opportunity to advance clean energy economy," *The Narwhal*, Feb. 10, 2016. https://thenarwhal.ca/trudeau-national-climate-meeting-seen-opportunity-advance-clean-energy-economy.
8. Stewart Elgie and Lorraine Mitchelmore, "Climate deal brings unlikely partnership between oil industry, environmentalist, first nations," *Huffpost*, Dec. 16, 2015. http://www.huffingtonpost.ca/stewart-e|gie/letter-to-trudeau-oil-environment-groups_b_8808172.html.
9. "Prime Minister Justin Trudeau speaks at the Smart Prosperity kick off in Vancouver," Mar. 1, 2016. https://www.youtube.com/watch?v=B-jjsqzqRVo.
10. There were many connections between Sustainable Prosperity and the Trudeau government, a relationship that seems to be central to Sustainable Prosperity's mission to bring market thinking into the government agency. Former EC deputy minister Alan Nymark, who died in 2016, was a Smart Prosperity member. Jane McDonald spent a year each as SP executive director and as director of policy for McKenna. SP policy director Michelle Brownlee had a ten-year career as a policy adviser at Natural Resources Canada and the Privy Council Office. SP senior research associate Tony Young joined the organization from Environment Canada, where he had been a director general in the Strategic Policy Branch, while senior research associate Meg Ogden had 15 years of policy research with various departments in the Canadian government.
11. Smart Prosperity Initiative, "New Thinking: Canada's Roadmap to Smart Prosperity," Mar. 2016. http://www.smartprosperity.ca/thinking/newthinking.
12. "Vancouver Declaration on Clean Growth and Climate Change," Vancouver, Prime Minister's Office, Mar. 3, 2016. https://pm.gc.ca/eng/news/2016/03/03/communique-canadas-first-ministers.
13. Jane Taber and Adrian Morrow, "Premiers agree on energy strategy with weakened climate change pledges," *Globe and Mail*, July 17, 2015. https://www.theglobeandmail.com/news/national/premiers-making-progress-on-national-energy-strategy-deal-could-be-signed-today/article25545448/.
14. Peter O'Neil, *op. cit.*
15. John Ibbitson, "Is Trudeau's national carbon tax even big enough to be worth it?" *Globe and Mail*, "Mar. 9, 2016. https://www.theglobeandmail.com/news/politics/globe-politics-insider/is-trudeaus-national-carbon-tax-even-big-enough-to-be-worth-it/article29100228/.
16. Shawn McCarthy and Robert Fife, "Ottawa may have to pay for carbon credits to meet climate targets," *Globe and Mail*, Dec. 8, 2016. https://www.theglobeandmail.com/news/politics/ottawa-may-have-to-pay-for-carbon-credits-to-meet-climate-targets/article33267047/.

17. James Wilt, "The carbon offset question," *The Narwhal*, Dec. 13, 2016. https://thenarwhal.ca/carbon-offset-question-will-canada-buy-its-way-climate-finish-line/.
18. "Canada's climate action is working, report to United Nations confirms," Government of Canada, Dec. 29, 2017. https://www.canada.ca/en/environment-climate-change/news/2017/12/canada_s_climateactionisworkingreporttounitednationsconfirms.html.
19. Peter Zimonjic and Susan Lunn, "Canada must reduce emissions from oilsands to meet climate goals," CBC News, Dec. 19.2017. http://www.cbc.ca/news/politics/oecd-canada-climate-report-card-1.4455379.
20. Mia Rabson, "Canadian emissions creep lower in 2016 but Paris targets still dubious," *Canadian Press*, Apr. 8, 2018.
21. John Geddes, "Building a consensus on climate change? Not so easy after all," *Maclean's*, Mar. 9, 2016. https://www.macleans.ca/politics/ottawa/building-a-consensus-on-climate-change-not-so-easy-after-all/.
22. Peter O'Neil, *op. cit.*
23. Jeff Lewis, "Suncor caps investments as pipeline projects stall," *Globe and Mail*, May 4, 2018, B1.
24. Jeff Lewis, "Enbridge Line 3 faces hurdle as US judge rejects route," *Globe and Mail*, Apr. 25, 2018, B1.
25. Logan Carroll, "How Enbridge helped write Minnesota pipeline laws aiding its Line 3 battle today," DeSmog, May 22, 2018. https://www.desmogblog.com/2018/05/22/how-enbridge-helped-write-minnesota-pipeline-laws-sf-90-line-3-battle.
26. "Nebraska OK's 'alternative route' for Keystone XL pipeline," CBC News, Nov. 20, 2017. http://www.cbc.ca/news/business/nebraska-keystone-1.4409960.
27. Mike De Souza, "Government insiders say Trans Mountain pipeline approval was rigged," *National Observer*, Apr. 26, 2018. https://www.nationalobserver.com/2018/04/24/kinder-morgan-opponents-suspected-trudeau-government-rigged-its-review-pipeline-federal.
28. Mychaylo Prystupa, "Federal ministers argue Trans Mountain expansion is necessary part of climate plan," *The Tyee*, Mar. 21, 2018. https://thetyee.ca/News/2018/03/21/Trans-Mountain-Climate-Plan/.
29. Andrew Nikiforuk, "Canada's dirty $20-billion pipeline bailout," *The Tyee*, May 29, 2018. https://thetyee.ca/Opinion/2018/05/29/Canada-Dirty-Pipeline-Bailout/.
30. Environment and Climate Change Canada, "Federal actions for a clean growth economy," Gatineau, QC, 2016, 3. https://www.canada.ca/content/dam/themes/environment/documents/weather1/20170119-en.pdf.
31. "Canada's clean growth century," Clean Growth Century Initiative, Dec. 5, 2016. http://policyoptions.irpp.org/magazines/december-2016/canadas-clean-growth-century/.
32. One exception is the Catherine Donnelly Foundation, which was established by the Sisters of Service, a Catholic women's religious order.
33. Quoted in Edouard Morena, "Climate philanthropy – the tyranny of the 2 per cent," *Alliance*, Aug. 2, 2016. http://www.alliancemagazine.org/analysis/tyranny-of-the-2-per-cent/.
34. "Economy and environment program framework," Ivey Foundation, 2015. http://www.ivey.org/wp-content/uploads/2016/04/IVEY-Economy-Environment-Program-Framework-2016.pdf.
35. Edouard Morena, "Follow the money," in Stefan Aykut, Jan Foyer and Edouard Morena, eds., *Globalising the Climate: COP21 and the Climatisation of Global Debates* (New York: Routledge, 2017), 97.
36. "Minister McKenna advances Canada's climate leadership at Paris One Planet Summit," Environment and Climate Change Canada, Dec. 13, 2017. https://www.canada.ca/en/environment-climate-change/news/2017/12/minister_mckennaadvancescanadasclimateleadershipatparisoneplanet0.html.
37. "Canada and World Bank Group to support the clean energy transition in developing countries and small island developing states," World Bank Group, Dec. 12, 2017. http://www.worldbank.org/en/news/press-release/2017/12/12/canada-and-the-world-

bank-group-to-support-the-clean-energy-transition-in-developing-countries-and-small-island-developing-states.

38. "Powering Past Coal Alliance: Declaration," Gov.UK, Apr. 9, 2018. https://assets. publishing.service.gov.uk/government/uploads/system/uploads/attachment_data/ file/700613/powering-past-coal-declaration.pdf.

39. "Canada and the United Kingdom team up with Bloomberg Philanthropies to support global efforts to phase out coal power," Environment and Climate Change Canada, Apr. 9, 2018. https://www.canada.ca/en/environment-climate-change/news/2018/04/ canada-and-the-united-kingdom-team-up-with-bloomberg-philanthropies-to-support-global-efforts-to-phase-out-coal-power.html.

40. Katie Fehrenbacher, "Breakthrough Energy Ventures unveils its 5-part investment strategy," Green Tech Media, Dec. 15, 2017. https://www.greentechmedia.com/articles/ read/breakthrough-energy-ventures-investment-strategy.

41. Oliver Milman, "Zuckerberg, Gates and other tech titans form clean energy investment coalition," *The Guardian*, Nov. 30, 2015. https://www.theguardian.com/ environment/2015/nov/30/bill-gates-breakthrough-energy-coalition-mark-zuckerberg-facebook-microsoft-amazon.

42. CIBC, "Energy," no date. http://www.cibcwm.com/cibc-eportal-web/portal/ wm?pageId=energy&language=en_CA.

43. Business Council of Canada, "Canada's oil sands: A vital national asset," Feb. 1, 2017. http://thebusinesscouncil.ca/publications/canadas-oil-sands-vital-national-resource/.

CHAPTER 14

1. "Latest CO$_2$ reading," Scripps Institution of Oceanography, Nov, 18, 2017. https:// scripps.ucsd.edu/programs/keelingcurve/wp-content/plugins/sio-bluemoon/graphs/ carbon dioxide_800k_zoom.png.

2. Nicola Jones, "How the world passed a carbon threshold and why it matters," *Yale Environment 360*, Jan. 26, 2017. https://e360.yale.edu/features/how-the-world-passed-a-carbon-threshold-400ppm-and-why-it-matters.

3. Matt McGrath, "Record surge in atmospheric CO$_2$ seen in 2016," BBC News, Oct. 30, 2017. http://www.bbc.com/news/science-environment-41778089.

4. "Greenhouse gas concentrations surge to new record," *World Meteorological Organization*, Oct. 30, 2017. https://public.wmo.int/en/media/press-release/greenhouse-gas-concentrations-surge-new-record.

5. "Global carbon dioxide emissions up about 2 per cent, scientists say," *Associated Press*, Nov. 13, 2017. https://www.cbsnews.com/news/carbon-dioxide-emissions-rise-globally/.

6. Nicola Jones, *op. cit.*

7. "350 PPM, not 450, should be our target," Stockholm Resilience Centre, 2008. http:// www.stockholmresilience.org/news--events/general-news/2008-06-23---350-ppm-not-450-should-be-our-target.html.

8. "350.org is building the global grassroots climate movement that can hold our leaders accountable to science and justice," *350.org.*, no date. https://350.org/about/

9. "Graphic: Carbon dioxide hits new high," Global Climate Change, NASA's Jet Propulsion Lab, California Institute of Technology, Nov. 13, 2017. https://climate.nasa. gov/climate_resources/7/.

10. For a history of the two-degree limit until 2014, see "Two degrees: The history of climate change's speed bump," *Carbon Brief*, Dec. 8, 2014. https://www.carbonbrief.org/ two-degrees-the-history-of-climate-changes-speed-limit.

11. "Statement of Dr. James Hansen, Director, NASA Goddard Institute for Space Studies," presented to United States Senate, June 23, 1988. https://climatechange.procon.org/ sourcefiles/1988_Hansen_Senate_Testimony.pdf.

12. William Stevens, "Earlier harm seen in global warming," *New York Times*, Oct. 17, 1990, A9.

13. "United Nations Framework Convention on Climate Change, Article 2," United Nations, 1992. http://unfccc.int/files/essential_background/background_publications_htmlpdf/application/pdf/conveng.pdf.

14. Malcolm Gladwell, *The Tipping Point* (New York: Little Brown, 2000).

15. Michael Ellison, "Perspectives on climate tipping points," American Geophysical Union, July 28, 2016. https://eos.org/editors-vox/perspectives-climate-tipping-points.

16. "Two degrees," *op. cit.*

17. Adrian Raftery *et. al.*, "Less than 2° C warming by 2100 unlikely," *Nature Climate Change*, 2017, Vol. 7, 637–641. https://www.nature.com/articles/nclimate3352.

18. Thorsten Mauritsen and Robert Pincus, "Committed warming inferred from observations," *Nature Climate Change*, 2017, Vol. 7, 652–655. https://www.nature.com/articles/nclimate3357.

19. Andrew Schurer *et al.*, "Importance of the pre-industrial baseline for likelihood of exceeding Paris goals," *Nature Climate Change*, 2017, Vol. 7, 563–567. https://www.nature.com/articles/nclimate3345.

20. "Only five years left before 1.5C carbon budget is blown," Carbon Brief, May 19, 2016. https://www.carbonbrief.org/analysis-only-five-years-left-before-one-point-five-c-budget-is-blown.

21. Mariette Le Roux, "'Carbon budget' may be bigger than thought: study," *Science X*, Sept. 19, 2017. https://phys.org/news/2017-09-carbon-bigger-thought.html.

22. Scott Johnson, "Possible good news about climate change leads to confusing coverage," *Arstechnica*, Sept. 22, 2017. https://arstechnica.com/science/2017/09/possible-good-news-about-climate-change-leads-to-confused-coverage/.

23. Sarah Kaplan, "Thousands of scientists issue bleak 'second notice' to humanity," *Washington Post*, Nov. 13, 2017. https://www.washingtonpost.com/news/speaking-of-science/wp/2017/11/13/thousands-of-scientists-issue-bleak-second-notice-to-humanity/.

24. International Energy Agency, "World Energy Outlook 2017," Paris, Nov. 14, 2017. https://www.iea.org/weo2017.

25. "Shell CEO: We can't ignore the importance of renewable energy," *Blue and Green Tomorrow*, June 4, 2015. http://blueandgreentomorrow.com/2015/06/04/shell-ceo-we-cant-ignore-the-importance-of-renewable-energy/.

26. John Ashton, "Shell's climate change strategy: narcissistic, paranoid and psychopathic," *Climate Change News*, Mar. 16, 2016. http://www.climatechangenews.com/2015/03/16/shells-climate-change-strategy-narcissistic-paranoid-and-psychopathic/.

27. *Ibid.*

28. Thomas Escritt, "Shell CEO expects no valuation hit from climate accord," *Reuters*, Nov. 26, 2016. https://www.reuters.com/article/us-shell-climatechange/shell-ceo-expects-no-valuation-hit-from-climate-accord-idUSKBN13L094.

29. Matthew Campbell, Rakteem Katakey and James Paton, "The future of Big Oil? At Shell, it's not oil," *Bloomberg*, July 19, 2016. https://www.bloomberg.com/news/articles/2016-07-20/the-future-of-big-oil-at-shell-it-s-not-oil.

30. Michael Crothers, "Decarbonization: Discovery, design and decisions shaping the 21st century economy," Speech at the 2016 Globe Conference, Vancouver, Mar. 3, 2016. https://www.shell.ca/en_ca/media/speeches/decarbonization.html.

31. "Natural gas," *David Suzuki Foundation*, no date. http://www.davidsuzuki.org/issues/climate-change/science/energy/natural-gas/.

32. Matthew Campbell *et al.*, *op cit.*

33. David Suzuki, "Natural gas is not a solution for climate change," July 19, 2011. http://www.davidsuzuki.org/blogs/science-matters/2011/07/natural-gas-is-not-a-solution-for-climate-change/.

34. Clifford Krauss, "Pressured, Shell vows to cut down on carbon," *New York Times*, Nov. 29, 2017, 4.

35. "Annual Report and Form 20-F 2016," Royal Dutch Shell, 2016, 6. http://reports.shell.com/annual-report/2016/.

36. *Ibid.*, 90.
37. Arthur Neslen, "Shell lobbied to undermine EU renewables targets, documents reveal," *The Guardian*, Apr. 27, 2015. http://www.theguardian.com/environment/2015/apr/27/shell-lobbied-to-undermine-eu-renewables-targets-documents-reveal.
38. Lauren Krugel, "Shell CEO pushes carbon capture and storage as way to combat climate change," *Canadian Press*, Nov. 9, 2015. http://www.ctvnews.ca/business/shell-ceo-pushes-carbon-capture-and-storage-as-way-to-combat-climate-change-1.2649877.
39. Clifford Krauss, *op. cit.*
40. Yadullah Hussain, "Shell says carbon storage without government help would be 'quite a challenge,'" *National Post*, Nov. 6, 2015. http://business.financialpost.com/news/energy/shell-says-carbon-storage-without-government-help-would-be-quite-a-challenge.
41. Lauren Krugel, *op. cit.*
42. "Energy transitions: Towards net zero emissions (NZE)" Royal Dutch Shell, no date. https://drive.google.com/file/d/0B_L1nw8WLu0Bbi1QWnJRcHlZblE/view. See also Georgie Johnson and Damian Kahya, "Leaked: the strategy behind Shell's low emissions PR push," *Greenpeace*, July 7, 2016. http://energydesk.greenpeace.org/2016/07/07/leaked-strategy-behind-shells-low-emissions-pr-push/; Mike Gaworecki, "Inside Shell's PR strategy to position itself as a 'net-zero emissions' leader," *DeSmog*, July 10, 2016. http://www.desmogblog.com/2016/07/10/inside-shell-s-pr-strategy-position-itself- net-zero-emissions-leader.
43. "A better life with a healthy planet: Pathways to net-zero emissions," Shell Global, May 2016, 2. http://www.shell.com/energy-and-innovation/the-energy-future/scenarios/a-better-life-with-a-healthy-planet.html/.
44. "Made-in-Canada clean tech receives a boost through Shell's Quest Climate Grant," *Canada Newswire*, Nov. 29, 2016. http://www.newswire.ca/news-releases/made-in-canada-clean-tech-receives-a-boost-through-shells-quest-climate-grant-603566566.html.
45. Dan Fumano, "'Timely' tax breaks hailed by LNG industry," *Vancouver Sun*, Mar. 23, 2018, A4.
46. Brent Jang, "LNG Canada sees Petronas as boost to 'dream team' following buy in," *Globe and Mail*, June 1, 2018, B3.
47. Kenneth Green and Niels Veldhuis, "B.C. LNG breakthrough — a flicker of hope in Canada's gloomy energy climate," *Vancouver Sun*, Mar. 22, 2018. http://vancouversun.com/opinion/op-ed/kenneth-green-and-niels-veldhuis-b-c-lng-breakthrough-a-flicker-of-hope-in-canadas-gloomy-energy-climate.
48. Donald Gutstein, "Follow the money, part 5 — The Tobacco Papers revisited," *Rabble.ca*, Apr. 24, 2014. http://rabble.ca/blogs/bloggers/donald-gutstein/2014/04/follow-money-part-5-tobacco-papers-revisited.
49. Kenneth Green, "Managing the risks of hydraulic fracturing," Fraser Institute, Dec. 2014, 24. https://www.fraserinstitute.org/sites/default/files/managing-the-risks-of-hydraulic-fracturing.pdf.
50. *Ibid.*, 8.
51. "Information circular — Proxy statement," NuVista Energy, Mar. 29, 2018. http://www.nuvistaenergy.com/files/4415/2511/0550/NuVista_MIC.pdf.
52. "Environmental impacts of shale gas extraction in Canada," Council of Canadian Academies, 2014. http://www.scienceadvice.ca/en/assessments/completed/shale-gas.aspx.
53. *Ibid.*, xv.
54. *Ibid.*, 215.
55. Carolyn Raffensperger, "The precautionary principle: A fact sheet," *Science and Environmental Health Network*, Mar. 1998. https://www.sehn.org/Volume_3-1.html.
56. "Environmental impacts of shale gas," *op. cit.*, xx.
57. Kenneth Green, *op. cit.*, 20.
58. "Environmental impacts of shale gas," *op. cit.*, xvii.
59. Kenneth Green, *op. cit.*, 11.

60. Radisav Vidic *et al.*, "Impact of shale gas development on regional water quality," *Science*, Vol. 340, May 17, 2013, 6.
61. "Management of wastes from the exploration, development and production of crude oil, natural gas and geothermal energy," Environmental Protection Agency, Report to Congress, Washington D.C., Dec. 1987.
62. John Chillibeck, "Fracking ban hurts economic opportunity," *St. John Telegraph and Journal*, Dec. 12, 2014, A1.
63. Kenneth Green and Taylor Jackson, "Managing the risks of hydraulic fracturing: An update," Fraser Institute, Oct. 29, 2015. https://www.fraserinstitute.org/studies/managing-the-risks-of-hydraulic-fracturing-an-update.pdf.
64. Philip Mirowski, *Never Let a Serious Crisis Go to Waste* (London: Verso, 2013), 338.
65. Hilda McKenzie and William Rees "An analysis of a brownlash report," *Ecological Economics*, Vol. 61, Iss. 2-3, Mar. 2007.
66. Ross McKitrick and Elmira Aliakbari, "Canada's air quality since 1970: An environmental success story," Fraser Institute, Apr. 2017, i. https://www.fraserinstitute.org/sites/default/files/canadas-air-quality-since-1970-an-environmental-success-story.pdf.

CHAPTER 15

1. For other instances of technological determinism in Canadian policy-making, see Donald Gutstein, *e.con: How the Internet Undermines Democracy* (Toronto: Stoddart, 1999), 69–74; Robert Babe, *Telecommunications in Canada* (Toronto: University of Toronto Press, 1990), 6–8.
2. "Pan-Canadian Framework on Clean Growth and Climate Change," Canada, 2016, 1. https://www.canada.ca/en/services/environment/weather/climatechange/pan-canadian-framework/climate-change-plan.html.
3. Ben Glasson, "Gentrifying climate change," M/C Journal, Vol. 15, No. 3, 2012. http://journal.media-culture.org.au/index.php/mcjournal/article/view/501.
4. Google the Clean Energy Canada website and the word inequality doesn't come up.
5. James Davies, Xiaojun Shi and John Whalley, "The possibilities for global inequality and poverty reduction using revenues from global carbon pricing," *Journal of Economic Inequality*, Vol. 12, 2014, 363–391.
6. William Rees, "Staving off the coming global collapse," *The Tyee*, July 17, 2018. https://thetyee.ca/Opinion/2017/07/17/Coming-Global-Collapse/.
7. "Economy is not science," David Suzuki, in *Surviving Progress*, Mathieu Roy and Harold Crooks, Cinemaginaire and Big Picture Media Corp., Sept. 2011. https://www.youtube.com/watch?v=4NiauhOCfsk.
8. Joseph Roberts, "Sustainable activism," a conversation with David Suzuki on his 75th birthday," *Common Ground*, Mar. 4, 2011. https://commonground.ca/sustainable-activism/.
9. David Suzuki with Faisal Moola, "Nature's bottom line," *Common Ground*, Mar. 11, 2010. https://commonground.ca/natures-bottom-line/.
10. Mike Moffat, "David Suzuki needs an economics refresher course," *Globe and Mail*, Oct. 10, 2012. https://www.theglobeandmail.com/report-on-business/economy/economy-lab/david-suzuki-needs-an-economics-refresher-course/article4602350/.
11. Stephen Gordon, "David Suzuki versus the economists," *Maclean's*, Oct 12, 2012. https://www.macleans.ca/economy/business/david-suzuki-versus-the-economists/.
12. Tristin Hopper, "'No way I'd share stage with Suzuki:' economist," *National Post*, Apr. 19, 2018, A1.
13. Bill Rees, "David Suzuki is right," *The Tyee*, May 4, 2018. https://thetyee.ca/Opinion/2018/05/04/David-Suzuki-Is-Right/.
14. Joseph Roberts, *op. cit.*
15. "The early Keeling curve," Scripps CO2 Program, 2007. https://web.archive.org/web/20070709142518/http://scrippscarbon dioxide.ucsd.edu/program_history/early_keeling_curve.html.

16. Indigenous is used instead of aboriginal, whose common dictionary meaning is "being the first of its kind present in a region and often primitive in comparison with more advanced types." [Webster's New Collegiate Dictionary]. Such a formulation obviously is not acceptable. Indigenous is more relevant in terms of its application to people: "having originated in and being produced, growing or living naturally in a particular region or environment." The use of the word environment is key.

17. Mitch Paquette, "Indigenous rights cut from Paris Agreement," Center for World Indigenous Studies, Jan. 13, 2016. https://intercontinentalcry.org/indigenous-rights-cut-from-paris-agreement-why-it-concerns-us-all/.

18. Article 7.5, *Paris Agreement*, United Nations, Dec. 12, 2015. https://unfccc.int/files/essential_background/convention/application/pdf/english_paris_agreement.pdf.

19. Mychaylo Prystupa, "'Trudeau fights to keep Indigenous rights in Paris climate deal," *National Observer*, Dec. 7, 2015. https://www.nationalobserver.com/2015/12/07/news/trudeau-fights-keep-indigenous-rights-paris-climate-deal.

20. Andrew Nikiforuk, "Megadams not clean or green, says expert," *The Tyee*, Jan. 23, 2018. https://thetyee.ca/News/2018/01/24/Megadams-Not-Clean-Green/.

21. Brandi Morin, "Where does Canada sit 10 years after the UN Declaration on the Rights of Indigenous Peoples?" CBC News, Sept. 13, 2017. http://www.cbc.ca/news/indigenous/where-does-canada-sit-10-years-after-undrip-1.4288480.

22. Hon. Lance Finch, "The duty to learn: Taking account of Indigenous legal orders in practice," Continuing Legal Education of BC, Indigenous legal orders and the common law conference, Nov, 15, 2012.

23. *Delgamuukw v. British Columbia*, [1997] 3 S.C.R. 1010 para. 147–48.

24. John Borrows, *Canada's Indigenous Constitution* (Toronto: University of Toronto Press, 2010). 23–24.

25. Louise Mandell, Chancellor's Speech to Vancouver Island University Convocation, Jan. 27, 2018. Personal Communication.

26. Lance Finch, *op. cit.*, para. 14.

27. Stephen Starr, "Norway's Sami communities fight to keep traditions," *The Irish Times*, Jan. 18, 2018, 8.

28. Sean Power, "Brazil: Ethanol interests on Guarani land," NACLA, no date. https://nacla.org/news/brazil-ethanol-interests-guaran%C3%AD-land.

29. Joao Ripper, "Death threat from Shell supplier on Brazilian tribe's land," *Survival*, May 31, 2011. https://www.survivalinternational.org/news/7329.

30. "Shell backs out of Brazil sugar-cane plans," *Agence France Presse*, June 13, 2012.

31. Chris Varcoe, "Alberta aims to repeat historic prices at new renewable power auctions," *Calgary Herald*, Feb. 6, 2018. http://calgaryherald.com/business/energy/varcoe-alberta-aims-to-repeat-historic-prices-at-new-renewable-power-auctions.

32. Ashish Kothari, Mari Margil and Shrishtee Bajpai, "Now rivers have the same legal status as people, we must uphold their rights," *The Guardian*, Apr. 21, 2017. https://www.theguardian.com/global-development-professionals-network/2017/apr/21/rivers-legal-human-rights-ganges-whanganui.

33. "Republic of Ecuador Constitution of 2008," Ecuador, National Assembly, Oct. 20, 2008. http://pdba.georgetown.edu/Constitutions/Ecuador/english08.html.

34. Natalia Greene, "The first successful case of the Rights of Nature implementation in Ecuador," Global Alliance for the Rights of Nature, 2018. https://therightsofnature.org/first-ron-case-ecuador/.

35. "Law of the Rights of Mother Earth," The Green Centre, no date. http://www.thegreencentre.co.uk/get-involved/law-of-the-rights-of-mother-earth.aspx.

36. John Vidal, "Bolivia enshrines natural world's rights with equal status for Mother Earth," *The Guardian*, Apr. 10, 2011. https://www.theguardian.com/environment/2011/apr/10/bolivia-enshrines-natural-worlds-rights.

37. Ashish Kothari *et al.*, *op. cit.*

38. "The Sami Parliament endorses the Universal Declaration of the Rights of Mother Earth," *Intercontinental Cry*, June 26, 2018. https://intercontinentalcry.org/the-sami-parliament-endorses-the-universal-declaration-of-the-rights-of-mother-earth/.
39. Adam Winkler, "'Corporations are people' is built on an incredible 19th-century lie," *The Atlantic*, Mar. 5, 2018. https://www.theatlantic.com/business/archive/2018/03/corporations-people-adam-winkler/554852/.
40. See Jane Mayer, *Dark Money* (New York: Doubleday, 2016), Chap. 9: "Money is speech: The long road to Citizens United."

CONCLUSION

1. Nina Chestney, "Carbon emissions hit new record in 2017," *Globe and Mail*, Mar. 23, 2018, B7.
2. "Another climate milestone on Mauna Loa," NOAA Research News, June 7, 2018. https://research.noaa.gov/article/ArtMID/587/ArticleID/2362/Another-climate-milestone-falls-at-NOAA%E2%80%99s-Mauna-Loa-observatory.
3. Jason Samenow, "Red-hot planet," *Washington Post*, July 5, 2018. https://www.washingtonpost.com/news/capital-weather-gang/wp/2018/07/03/hot-planet-all-time-heat-records-have-been-set-all-over-the-world-in-last-week/.
4. Eric Schmidt, "Electric vehicles sales update Q1 2018, Canada," June 8, 2018. https://www.fleetcarma.com/electric-vehicles-sales-update-q1-2018-canada/.
5. Timothy Cain, "Canada's 30 best-selling vehicles overall, one quarter through 2018," *Autofocus*, Apr. 9, 2018. http://www.autofocus.ca/news-events/canadian-car-and-truck-sales/canada-s-30-best-selling-vehicles-overall-one-quarter-through-2018.
6. "Perceptions of carbon pricing in Canada," *Abacus Data*, Feb. 2018. https://ecofiscal.ca/wp-content/uploads/2018/04/Ecosfiscal_Polling_February2018_FINAL_RELEASE.pdf.
7. "Jason Kenney to climate change denier: 'the climate's been changing since the beginning of time,'" *Press Progress*, Jan. 3, 2017. http://pressprogress.ca/jason_kenney_to_climate_change_denier_the_climate_been_changing_since_the_beginning_of_time/.
8. "2018 Ontario Provincial Election: The Leaders' Final TV Debate," May 27, 2018. https://www.youtube.com/watch?v=k8GPir1uuk4.
9. Estefania Duran and Marcella Bernardo, "Burnaby's option to fight pipeline expansion now very limited, says political scientist," *News1130*, Aug. 23, 2018. https://www.news1130.com/2018/08/23/burnaby-trans-mountain-pipeline-expansion/.

INDEX